AF386727

Synthesis Lectures on Mathematics & Statistics

Series Editor

Steven G. Krantz, Department of Mathematics, Washington University, Saint Louis, USA

This series includes titles in applied mathematics and statistics for cross-disciplinary STEM professionals, educators, researchers, and students. The series focuses on new and traditional techniques to develop mathematical knowledge and skills, an understanding of core mathematical reasoning, and the ability to utilize data in specific applications.

Jennifer J. Edmond • Jack E. Graver

The Linear Function and Euclidean Geometry

Springer

Jennifer J. Edmond
Private Tutor
Baldwinsville, NY, USA

Jack E. Graver
Department of Mathematics
Syracuse University
Syracuse, NY, USA

ISSN 1938-1743   ISSN 1938-1751   (electronic)
Synthesis Lectures on Mathematics & Statistics
ISBN 978-3-032-22711-9   ISBN 978-3-032-22712-6   (eBook)
https://doi.org/10.1007/978-3-032-22712-6

This Springer imprint is published by the registered company Springer Nature Switzerland AG
The registered company address is: Gewerbestrasse 11, 6330 Cham, Switzerland

If disposing of this product, please recycle the paper.

Preface

This book is intended to serve as a bridge between high school and college mathematics. Its central theme, as the title suggests, is the linear function. We have designed the book to highlight the relationships among topics from the high school curriculum, including material commonly covered in a 12th year non-calculus course, as well as in first and second year college courses. The book is divided into two parts, devoted, respectively, to linear functions of a real variable and to linear functions of a two-dimensional vector variable. Throughout, we emphasize the visualization of these functions as transformations of one- and two-dimensional Euclidean space.

One of the common concerns expressed by colleges is that many incoming freshmen do not possess algebra as a usable tool. In this text, algebra is presented as an essential foundation for further study in mathematics. Algebraic skills are strengthened through consistent use and by demonstrating their application in a wide variety of contexts.

As educators ourselves, we have observed that many students view mathematics as a collection of isolated subjects, each with its own methods, techniques, and results. This book takes a holistic approach to challenge that perception. The topics, examples, and problems have been carefully chosen to encourage the integration of major ideas from the high school and early college mathematics curriculum.

Another feature of this book is that it includes proofs. Not the formal "two column proofs" often associated with high school geometry, but the more informal, but nevertheless rigorous, step-by-step explanation of why what we believe to be true is indeed true.

In addition to its use in non-calculus courses, this book may also serve in the training of middle school and high school teachers, with the aim of deepening their understanding of the algebra and geometry that underlie the secondary school curriculum. We also hope that some high school students with a special interest in mathematics, as well as teachers, will choose to read this book simply because they find it interesting.

This book contains relatively few routine exercises. Readers are encouraged to create their own routine and non-routine examples. Central to the philosophy of this book is the idea that asking questions is just as important to mathematics as answering them.

Part I, The Real Linear Function, forms the core of the book and can be used as a stand-alone unit in either a high school or college-level course. The first chapter introduces our approach to real linear functions by interpreting them as transformations of the number line. The geometric consequences of this interpretation are the focus of Chap. 2.

Our geometric understanding of linear functions allows us to give a very intuitive derivation of the formula for the nth term of a solution to a linear difference equation. This derivation appears in Chap. 3, where linear difference equations are studied in detail. In that chapter, it becomes clear that iterating a linear function naturally leads to the study of exponential functions. One of the most useful and interesting applications of these exponential functions is to the mathematics of personal finance: compound interest, annuities, and loans. These topics are explored in Chap. 4.

Chapter 5 presents a variety of interesting applications of linear functions in business. Quadratic functions arise naturally in this context and enable us to review some of the algebra of quadratics. Also, several fundamental applications of calculus arise. However, instead of using derivatives, we introduce marginal functions: if $P(x)$ is the profit when producing x items, then $P^*(x)$ is defined as the marginal profit, the change in profit when one more item is produced. As demonstrated in Chap. 5, the marginal function serves as a powerful and intuitive pre-calculus substitute for the derivative.

Part II, The Two-Dimensional Linear Function, expands the same mathematical concepts and tools in Part I to two dimensions.

There are many people who must be thanked for their support during the writing of this book. First among them are our spouses, Yana Graver and Stephen Edmond. We offer special thanks to Stephen for his careful reading of the manuscript and his many helpful suggestions.

Some of the material here was designed for, and first used in, a New York State-funded mathematics camp for high school students which was held at Syracuse University during the summer of 1992. Sixteen talented young people from all over New York State spent three weeks at Syracuse University, exploring a wide variety of mathematical topics.

An expanded set of lectures on the linear function was then presented to a group of high school mathematics teachers in the program $MTRC^3$ (Mathematics Teacher/Researchers Collaborating for Collaboration in the Classroom). This program was funded by a National Science Foundation grant for four year period starting in the Spring of 1993. Many of the ideas developed in these workshops, were then published: A geometric approach to linear functions, The College Mathematics Journal, Vol. 26, No. 5, 1995, pp. 389–394 and then greatly expanded in Chaps. 1 and 2 here.

Baldwinsville, NY, USA
Syracuse, NY, USA

Jennifer J. Edmond
Jack E. Graver

Contents

Part I

The Real Linear Function

Introduction

1.1 Visualizing Linear Functions

A *real linear function* is defined as $f(x) = ax + b$, where a, b, and x are real numbers with $a \neq 0$. As is usual, a will be called the *slope* of $f(x) = ax + b$. Although constant functions of the form $f(x) = b$ (where the slope $a = 0$) produce straight-line graphs and are sometimes categorized as linear functions, we will adopt an algebraic perspective. In this context, the use the term "linear function" to refer specifically to polynomial functions of degree one, thereby excluding constant functions. These linear functions map real numbers to real numbers, and we conceptualize the real numbers as forming the one-dimensional Euclidean geometry or the *Euclidean line*, equipped with a coordinate system. The *distance* between any two real numbers x and y is given by the absolute value of their difference, $|x - y|$.

We often think of a linear function as a rule that assigns each point on one copy of the number line (the x-axis) to a point on another copy (the y-axis), and we visualize this relationship using a graph. In this book, we study real linear functions as transformations of the number line itself:

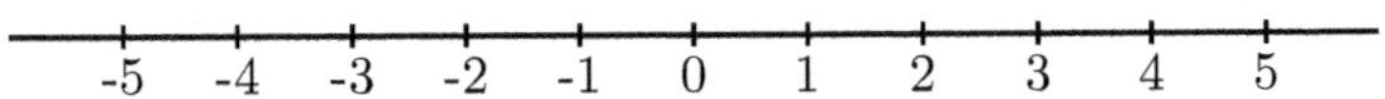

The feature of the number line that is critical to our discussion is that the scale is a *linear scale*, that is the distance on the line between any two numbers is proportional to the absolute value of the difference of those two numbers. So, for instance, 2 and 4 are the same distance apart as $\frac{3}{2}$ and $-\frac{1}{2}$ since $|2 - 4| = 2$ and $|\frac{3}{2} - (-\frac{1}{2})| = 2$.

We start our investigation by considering the function $f(x) = \frac{4}{5}x + 1$. We begin by drawing a picture of this function as a map from one copy of the Euclidean line onto a second copy of the Euclidean line (drawn above the first line):

J. J. Edmond, J. E. Graver, *The Linear Function and Euclidean Geometry*, Synthesis Lectures on Mathematics & Statistics, https://doi.org/10.1007/978-3-032-22712-6_1

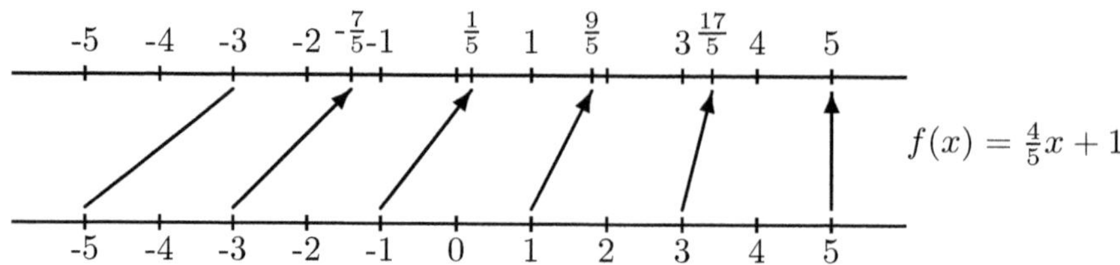

A yet simpler picture is obtained with just one copy of the number line by writing the function value, $f(x)$, directly above the x on the line:

We call this method of picturing a function the *one-line (two-scale) graph* of the function. Using this graph, we interpret f as a transformation of the line: the number 0 is mapped to 1, hence we think of 0 as being shifted to 1; similarly, -5 is shifted to -3, 2 to $\frac{13}{5}$, etc.

Observe that the set of function values form a second linear scale for the line:

Note however, that these two linear scales measure distances differently. For example, the points labeled by 2 and 4 on the original scale are a distance 2 apart on that scale. After applying the linear function they are labeled by $\frac{13}{5}$ and $\frac{21}{5}$ and are a distance $\frac{8}{5}$ apart on the second scale. This gives us a ratio of 1 to $\frac{4}{5}$. In fact, the ratio of the distances in the two scales is the same for all pairs of points: given any two points on the line, the distance between them, as measured by the top scale, is $\frac{4}{5}$ of the distance between them, as measured by the original scale on the bottom. Thus, the function f effectively reduces all distances by a factor of $\frac{4}{5}$.

Another feature of this one line graph of f, is that the two scales seem to be getting closer together as one continues to the right. Shifting our "viewing window" to the right, we get:

From this viewpoint, we can see a very nice geometric description for f: it is the contraction by a factor of $\frac{4}{5}$ about the point 5 (the center of the contraction). Still another interesting property is that the one-line graph of a linear function's inverse is obtained by simply interchanging the scales:

$$f^{-1}(x) = \tfrac{5}{4}x - \tfrac{5}{4}$$

For the reader unfamiliar with the concept of the inverse of a function, we will define it intuitively here for any function mapping the real line to the real line. Thinking of a function f geometrically as moving the points on the real line to other positions, we think of the inverse as the function from the real line to the real line that puts each point back in its original position. If such a function exists, we denote the inverse of f by f^{-1}. One of the reasons f^{-1} may not exist is that a function may map multiple points onto the same point. For example, if we consider $f(x) = x^2$ then both 2 and -2 are mapped onto 4 since $f(2) = f(-2) = 4$. When this happens there is no way to map 4 back to its original position since 4 has two different "original positions", 2 and -2, so there is no way to know which is the original starting position. We say that a function $f(x)$ is *one-to-one* if distinct points are mapped onto distinct points: if for all x and y with $x \neq y$, $f(x) \neq f(y)$.

Another reason that a function f from the real numbers to real numbers may not have an inverse is that it may not be onto. A function f from the real line to the real line is *onto*, if for every real number x, there is a real number y so that $f(y) = x$. For example, the function $f(x) = x^2$ is not onto: there is no input x that is mapped onto -1 so an inverse would have nothing to map -1 to. Every number is already mapped to some number different from -1. There is a general theorem in set theory that you may have already encountered: a function f has an inverse f^{-1} if and only if it is both one-to-one and onto. It is easy to show that a linear function $f(x) = ax + b$ is one to one and onto. And it is also straight forward to compute its inverse: just set $ax + b$ equal to y and solve for x:

$$ax + b = y$$
$$ax = y - b$$
$$x = \frac{y - b}{a}$$
$$x = \frac{1}{a}y - \frac{b}{a}$$

Theorem 1.1 *The inverse of the linear function $f(x) = ax + b$ is $f^{-1}(x) = \frac{1}{a}x - \frac{b}{a}$*

Note, the constant functions (slope $a = 0$) are not one-to-one, are not onto and do not have inverses since $\frac{1}{a}$ is undefined.

Exercise 1.1 Let $f(x) = \frac{4}{5}x + 1$.

a. For several values, check that $f^{-1}(x) = \frac{5}{4}x - \frac{5}{4}$.
b. What is the inverse of $f^{-1}(x) = \frac{5}{4}x - \frac{5}{4}$?
c. Give a geometric description of the action of $f^{-1}(x) = \frac{5}{4}x - \frac{5}{4}$ on the line.

The next exercise stems from two common examples of one-line graphs: a thermometer with both Fahrenheit and Celsius scales, and a meter stick with centimeters on one side and inches on the other. The dual-scale thermometer can be read in either direction— Celsius to Fahrenheit or Fahrenheit to Celsius—resulting in two linear functions that are inverses of each other. Similarly, the meter stick produces a pair of functions. However, on a standard meter stick, each scale is read from left to right when it is placed on top. While the standard meter stick provides a linear function, it does not convert centimeters to inches; only a converting meter stick does that. Before proceeding with this exercise, we need to prove an important fact about linear functions: they are uniquely determined by any two distinct function values: the two-point theorem.

Theorem 1.2 *Let $r \neq s$ be two distinct real numbers, let $f(x) = ax + b$ and $g(x) = a'x + b'$ and assume that $f(r) = g(r)$ and $f(s) = g(s)$; then $f = g$.*

Proof To prove that $f = g$, we will show that $a = a'$ and $b = b'$.
Since $f(r) = g(r)$ and $f(s) = g(s)$, we have:

$$ar + b = a'r + b' \quad \text{and} \quad as + b = a's + b'.$$

Equivalently we have:

$$r(a - a') + (b - b') = 0 \quad \text{and} \quad s(a - a') + (b - b') = 0.$$

$$r(a - a') + (b - b') = s(a - a') + (b - b') = 0,$$

$$r(a - a') - s(a - a') = (r - s)(a - a') = 0.$$

Now, since $r - s \neq 0$, it follows that $a - a' = 0$ and so $a = a'$. Since $r \neq s$ one of r and s is not 0, say r. Then $r(a - a') + (b - b') = 0$ becomes $(b - b') = 0$ or $b = b'$.
Therefore, $f(x) = ax + b = a'x + b' = g(x)$ for all x, so $f = g$. $\qquad\square$

Exercise 1.2 Consider the thermometer and the meter stick described above and pictured below.

a. Find the linear function which converts the Fahrenheit measure of the temperature to its Centigrade measure.
b. Find the linear function which converts the Centigrade measure of the temperature to its Fahrenheit measure.
c. Find the temperature which has the same measure in the Fahrenheit and Centigrade scales.
d. Find the linear function which has the standard meter stick as pictured below as its one-line graph.
e. Find the fixed point of the standard meter stick.
f. Find the function which converts centimeters to inches and its inverse.

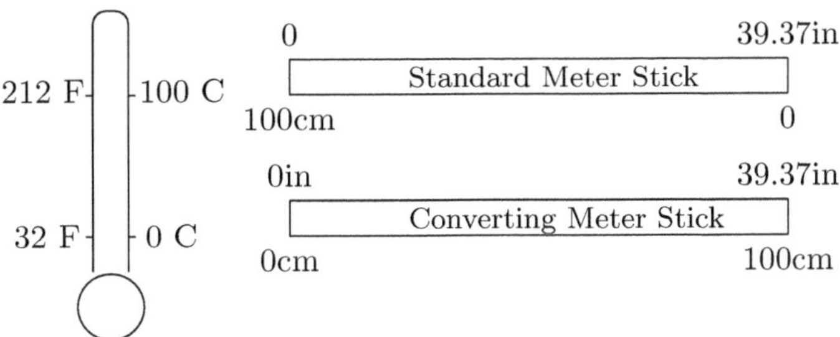

The examples discussed in this section are representative in that every real linear function has a clear geometric interpretation. It is this geometric perspective on real linear functions that we wish to explore. To proceed, we need a precise definition of the geometry of the line and of the transformations that act on it. That is the task that we turn to next.

1.2   The Real Linear Functions

In a Euclidean space of any dimension, distance is used to define the *Euclidean transformations* of that space. Hence, we introduce a coordinate system to our Euclidean spaces. For the 1-dimensional Euclidean space we have the real number line we have been using; for the two-dimensional Euclidean space we have the usual coordinate system for graphing functions. Since the definitions of Euclidean transformations are the same in all dimensions, we give the definitions in the more general setting. A mapping of a Euclidean space (of any dimension) onto itself is called a *similarity* if all distances between pairs of points are altered (magnified or contracted) by the same multiplicative factor, a positive real number, called the *magnification* of the similarity. We stated above, that $f(x) = \frac{4}{5}x + 1$ shrinks all distances by $\frac{4}{5}$. We can verify by direct computation that f is indeed a similarity of the line with magnification $\frac{4}{5}$. Let r and s be any two real numbers, then:

$$|f(r) - f(s)| = \left| \left(\frac{4}{5}r + 1 \right) - \left(\frac{4}{5}s + 1 \right) \right|$$

$$= \left| \frac{4}{5}r - \frac{4}{5}s \right|$$

$$= \left| \frac{4}{5}(r - s) \right|$$

$$= \frac{4}{5}|r - s|.$$

Lemma 1.1 *Let $f(x) = ax + b$ be a linear function. Then f is a similarity with magnification $|a|$.*

Proof We must show that f satisfies the definition of a similarity: Specifically, we must show that there is a positive real number a so that, for any two points r and s, the distance between $f(r)$ and $f(s)$ is a times the distance between r and s. By direct computation:

$$|f(r) - f(s)| = |(ar - b) - (as - b)|,$$

$$= |ar - as|,$$

$$= |a(r - s)|,$$

$$= |a||r - s|$$

Hence, f is a similarity with magnification $|a|$.   □

This lemma leads to a natural question: "Is every similarity of the Euclidean line given by a linear function?" After an exploration of the geometry of the line in the next chapter, we will confirm that the answer is "Yes".

If the magnification of a similarity is 1 ($a = \pm 1$), the similarity is called a *congruence* or *isometry*. As these terms indicate, a similarity will map a figure onto a *similar* figure and a congruence will map a figure onto a *congruent* figure. In fact, these transformations enable us to give formal definitions for the similarity and congruence of arbitrary sets. The set T in a Euclidean space is *similar* to the set S if there is a similarity f so that $T = f(S)$; the set T in a Euclidean space is *congruent* to the set S if there is a congruence f so that $T = f(S)$. In a later section, we will prove that the inverse of a similarity always exists and is a similarity and that the inverse of a congruence always exists and is a congruence. Hence, we may simply say that the sets S and T are similar (or congruent) if there is a similarity (or congruence) mapping one onto the other.

We have just explored the significance of the absolute value of the slope of a linear function; what about the significance of the "sign" of the slope? To answer this question we need a few additional definitions. We say that the standard scale or standard coordinate system for the real line is *positively oriented (or oriented to the right)* that is *for any two real numbers r and s, if r > s, then r is to the right of s on the number line.*

Now consider a linear function $f(x) = ax + b$ and let r and s be any two real numbers and assume that $r > s$. Then $f(r) - f(s) = (ar + b) - (as + b) = a(r - s)$.
Since $(r - s) > 0$, $f(r) > f(s)$ when a is positive and $f(s) > f(r)$ when a is negative. We conclude that the scale for the line given by a linear function f is positively oriented when the slope is positive. When the slope is negative, the numbers on the scale given by f are increasing as we move to the left, in the negative direction. In this case, we say the scale given by f is *negatively oriented (or oriented to the left)*. Since, in this case, the scale given by f is negatively oriented and we say that f *reverses orientation*; while in the case $a > 0$, the scale given by f is positively oriented and we say that f *preserves orientation*.
We combine this result with Lemma 1.1 to get:

Theorem 1.3 *Consider the linear function $f(x) = ax + b$; then:*

(1) *f is a similarity with magnification $|a|$;*
(2) *f preserves orientation if $a > 0$ and reverses orientation if $a < 0$.*

Exercise 1.3

a. Draw the one-line graph of $f(x) = x - 3$ and give a geometric description of this transformation of the line.
b. Draw the one-line graph of $g(x) = -x + 3$ and give a geometric description of this transformation of the line.
c. Verify that the functions f and g are congruences of the Euclidean line.

It follows from Theorem 1.3 that linear functions of the form $f(x) = x + b$ and $g(x) = -x + b$ are congruences of the Euclidean line. Geometrically, $f(x) = x + b$ moves every point on the line a distance $|b|$ to the right, when $b > 0$, and to the left, when $b < 0$. When $b = 0$, $f(x) = x$ is the *identity transformation*. We call a linear function of the form $f(x) = x + b$ a *translation*. It is convenient to adopt the notation $t_{[b]}(x) = x + b$ and the convention that $t_{[b]}$ is the translation by b units to the right, understanding a translation to the "right" by a negative number to be a translation to the left. For example:
Translation by 1 to the right: $t_{[1]}(x) = x + 1$

<table>
<tr><td>-4</td><td>-3</td><td>-2</td><td>-1</td><td>0</td><td>1</td><td>2</td><td>3</td><td>4</td><td>5</td><td>6</td><td>$t_{[1]}(x) = x + 1$</td></tr>
<tr><td>-5</td><td>-4</td><td>-3</td><td>-2</td><td>-1</td><td>0</td><td>1</td><td>2</td><td>3</td><td>4</td><td>5</td><td>x</td></tr>
</table>

Translation by $\frac{5}{4}$ to the left: $t_{[-\frac{5}{4}]}(x) = x - \frac{5}{4}$

$$\begin{array}{ccccccccccc}
-\frac{25}{4} & -\frac{21}{4} & -\frac{17}{4} & -\frac{13}{4} & -\frac{9}{4} & -\frac{5}{4} & -\frac{1}{4} & \frac{3}{4} & \frac{7}{4} & \frac{11}{4} & \frac{15}{4} \\[4pt]
-5 & -4 & -3 & -2 & -1 & 0 & 1 & 2 & 3 & 4 & 5
\end{array} \qquad \begin{array}{l} t_{[-\frac{5}{4}]}(x) = x - \frac{5}{4} \\[4pt] x \end{array}$$

In this notation, the *identity* function, the function which maps each point onto itself, is the translation by 0: $t_{[0]}(x) = x$.

Geometrically, the linear functions of the form $g(x) = -x + b$ are a bit more complicated. Note that g interchanges 0 and b: $g(0) = b$ and $g(b) = 0$. Furthermore, it maps $\frac{b}{2}$ onto itself: $g(\frac{b}{2}) = -\frac{b}{2} + b = \frac{b}{2}$. Consider the example $f(x) = -x + 3$. This function maps $\frac{3}{2}$ onto itself and its one-line graph is:

$$\begin{array}{cccccccccccc}
8 & 7 & 6 & 5 & 4 & 3 & 2 & \overset{3}{\underset{2}{}} & 1 & 0 & -1 & -2 \\[4pt]
-5 & -4 & -3 & -2 & -1 & 0 & 1 & \underset{\frac{3}{2}}{} & 2 & 3 & 4 & 5
\end{array} \qquad \begin{array}{l} f(x) = -x + 3 \\[4pt] x \end{array}$$

In this example, each point is reflected through the fixed point $\frac{3}{2}$: 1 is mapped to 2, and 2 is mapped to 1; 0 is mapped to 3, and 3 is mapped to 0.

More generally, for any number w, the point $\frac{3}{2} + w$ is mapped to $\frac{3}{2} - w$, and $\frac{3}{2} - w$ is mapped to $\frac{3}{2} + w$. We say this reflection is centered at $\frac{3}{2}$.

In general, a function of the form $f(x) = -x + b$ reflects all points across the fixed point $\frac{b}{2}$. We call this point the *center* of the function. Specifically,

$$\left(\frac{b}{2} + w\right) \text{ is mapped to } \left(\frac{b}{2} - w\right), \text{ and } \left(\frac{b}{2} - w\right) \text{ is mapped to } \left(\frac{b}{2} + w\right),$$

as you will now verify.

Exercise 1.4 Consider the linear function $g(x) = -x + b$ and show that g interchanges points equidistant from $\frac{b}{2}$. Specifically, show that g leaves $\frac{b}{2}$ fixed and that, for all $w > 0$, g interchanges the points $\frac{b}{2} + w$ and $\frac{b}{2} - w$.

We call a linear function of the form $g(x) = -x + b$ a *reflection*, specifically, *the reflection about* $\frac{b}{2}$. Replacing b by $2c$ in the formula, we adopt the notation $r_{[c]}(x) = -x + 2c$ and the terminology $r_{[c]}$ is the *reflection with center c*.

Exercise 1.5

a. Give the linear function for the translation by $\frac{5}{2}$ units to the right.
b. Give the linear function for the reflection about $\frac{5}{2}$.
c. Give the inverses of these two functions and give their geometrical descriptions.

Lemma 1.2

(1) *The inverse of the translation $t_{[b]}(x)$ is the translation $t_{[-b]}(x)$.*
(2) *The inverse of the reflection $r_{[c]}(x)$ is itself, $r_{[c]}(x)$.*

Exercise 1.6 Prove this lemma.

It is natural to ask whether translations and reflections are the only congruences of the Euclidean line. By Lemma 1.1, since translations and reflections are the only linear functions with $|a| = 1$, we know that they are the only congruences that can be represented by a linear function.

However, until we prove that *all* congruences are given by a linear function, we cannot rule out the possibility that other congruences exist. At this stage, we can only say that "we haven't found any others"—which is hardly a convincing proof!

Constructing a rigorous proof that translations and reflections are the only congruences of the Euclidean line is one of the central goals of the next chapter.

1.3 Compositions and Inverses

If $f(x)$ and $g(x)$ are any two functions, we may apply one and then the other to get a new function called the *composition* of f followed by g and denoted by $g \circ f$:

$$(g \circ f)(x) = g(f(x)) \text{ for each real number } x.$$

It is important to note that, in general, the order of composition matters. That is, the composition of f followed by g is not the same as the composition of g followed by f: Equivalently put, compositions are not commutative: $g \circ f$ is not the same as $f \circ g$.

For example, let $f(x) = \frac{4}{5}x + 1$ and $g(x) = \frac{3}{2}x - 1$. Then:

$$(g \circ f)(x) = g(f(x)) = \frac{3}{2}(f(x)) - 1 = \frac{3}{2}(\frac{4}{5}x + 1) - 1 = \frac{6}{5}x + \frac{1}{2}; \text{ while}$$
$$(f \circ g)(x) = f(g(x)) = \frac{4}{5}(g(x)) + 1 = \frac{4}{5}(\frac{3}{2}x - 1) + 1 = \frac{6}{5}x + \frac{1}{5}.$$

The fact that in this example, $(g \circ f)(x)$ and $(f \circ g)(x)$ are both linear functions is no accident. Let $f(x) = ax + b$ and $g(x) = a'x + b'$ denote arbitrary linear functions. Then $(g \circ f)(x) = a'(ax + b) + b' = (a'a)x + (a'b + b')$, which is a linear function. Since we will want to refer to this general fact later on, we will formalize it as a lemma:

Lemma 1.3 *Let $f = ax + b$ and $g = a'x + b'$ be two linear functions. Then $g \circ f$ is the linear function $(a'a)x + (a'b + b')$.*

Theorem 1.4 *Let $t_{[b]}$ and $t_{[b']}$ be two translations and let $r_{[c]}$ and $r_{[c']}$ be two distinct reflections ($c \neq c'$). Then*

(1) $t_{[b]} \circ t_{[b']} = t_{[b+b']}$.
(2) $r_{[c]} \circ r_{[c']} = t_{[2(c-c')]}$.
(3) $t_{[b]} \circ r_{[c]} = r_{[c+\frac{b}{2}]}$.
(4) $r_{[c]} \circ t_{[b]} = r_{[c-\frac{b}{2}]}$.

Proof By direct computation:

(1) $t_{[b]} \circ t_{[b']}(x) = (x + b') + b = x + (b + b') = t_{[b+b']}(x)$.
(2) $r_{[c]} \circ r_{[c']}(x) = -(-x + 2c') + 2c = x + 2(c - c') = t_{[2(c-c')]}(x)$.
(3) $t_{[b]} \circ r_{[c]}(x) = (-x + 2c) + b = -x + 2(c + \frac{b}{2}) = r_{[c+\frac{b}{2}]}(x)$.
(4) $r_{[c]} \circ t_{[b]}(x) = -(x + b) + 2c = -x + 2(c - \frac{b}{2}) = r_{[c-\frac{b}{2}]}(x)$.

$\square$

Now that we have the necessary tools, we can formalize the general concept of the inverse of a function. Intuitively, we viewed a function f as a geometric transformation of the line, and described its inverse, f^{-1}, as the transformation that returns each point to its original position. We can now restate this idea using the composition of functions: A function g is called the *inverse* of a function f if their composition returns every input to itself. That is,

$$(g \circ f)(x) = x \quad \text{for every } x.$$

In this case, the composition $g \circ f$ is the *identity function*: the function that leaves every point unchanged.

Exercise 1.7 Show that the inverse of a translation is a translation and that the inverse of a reflection is a reflection. Specifically, show both geometrically and algebraically that:

$$t_{[b]}^{-1} = t_{[-b]} \quad \text{and} \quad r_{[c]}^{-1} = r_{[c]}.$$

Exercise 1.8 Find all linear functions which are their own inverses.

1.4   The Slope-Center Form

By a *fixed point* for a function $f(x)$, we mean a real number c so that $f(c) = c$. Since translations other than the identity shift every point, they have no fixed points. The surprising fact is that every linear function that is not a translation must have a fixed point! If the linear function is given by $f(x) = ax + b$, the question as to whether it has fixed point becomes: does the equation $x = ax + b$ have a solution? The answer is yes, if either $a \neq 1$, or if $a = 1$ and $b = 0$. These observations are recorded in the next lemma and a formal proof follows.

Lemma 1.4 *Let $f(x) = ax + b$.*

(1) *If $a = 1$ and $b = 0$, then f is the identity function and all points are fixed.*
(2) *If $a = 1$ and $b \neq 0$, then f is a translation different from the identity and has no fixed points.*
(3) *If $a \neq 1$, then f is not a translation and $\frac{b}{1-a}$ is its unique fixed point.*

Proof The real number x is a fixed point for f if and only if $x = ax + b$, or equivalently if and only if $(1 - a)x = b$.

1. If $a = 1$ and $b = 0$, $(1 - a)x = 0 = b$, for all x, and every point is fixed. In this case, $f(x) = x$ which is the identity function.
2. If $a = 1$ and $b \neq 0$, $(1 - a)x = 0 \neq b$, for all x, and no point is fixed. In this case, $f(x) = x + b$ which is a translation different from the identity.
3. If $a \neq 1$, then f is not a translation. In this case, $(1 - a)x = b$ if and only if $x = \frac{b}{1-a}$.

$\square$

We call the fixed point of a linear function its *center*. A non-translation $f(x) = ax + b$ can be rewritten in *slope-center* form:

$$f(x) = ax + (1 - a)c, \quad \text{where } (1 - a)c = b \text{ and } c = \frac{b}{1 - a} \text{ is the center of } f.$$

In general, this form is useful for understanding the geometric properties of the function. Occasionally, a particular linear function has a natural or intuitive expression in slope-center form.

For example, consider a familiar linear function to many students: their final course average as a function of their (yet to be taken) final exam score. The center of this function is the student's current average before the final exam. If Steve currently has a 79% average in his course work and he gets a 79% on his final exam, his final average will be 79%. Thus, this function is naturally expressed in the form $f(x) = ax + (1 - a)c$, where:

- x is the score on the final exam (as a percent),
- c is the current average (as a percent),
- a is the weight of the final exam (as a decimal).

Suppose Steve's current average is 79%, and the final exam counts for 40% of the course grade. Then the function giving his final course average based on his exam score x is:

$$f(x) = 0.4x + (1 - 0.4)(0.79) = 0.4x + 0.474.$$

So, for example:

- If Steve scores 0% on the final, his final average is $f(0) = 0.4(0) + 0.474 = 47.4\%$.
- If he scores 50%, his final average is $f(50) = 0.4(50) + 0.474 = 67.4\%$.
- If he scores 79%, his final average is $f(79) = 0.4(79) + 0.474 = 79\%$: the center of the function.

By evaluating $f(100)$, we find that the best grade Steve can earn in the course is:

$$f(100) = 0.4(100) + 0.474 = 87.4\%.$$

The inverse function $g(x) = 2.5x - 1.185$ is also useful. It answers the question: what score on the final exam is needed to achieve a desired final course average?

For example, to earn at least a B− (80%), we compute:

$$g(0.80) = 2.5(0.80) - 1.185 = 0.815,$$

so you would need to score at least 81.5% on the final exam to earn a B− in the course. Similarly, to finish the course with an 85% average, you would need to score:

$$g(0.85) = 2.5(0.85) - 1.185 = 0.94,$$

that is, 94% on the final exam.

Exercise 1.9 Suppose that your present average is 86% and that the final exam accounts for 25% of your grade.

a. Describe your final average as a function of your (yet to be taken) final exam.
b. What is the highest grade that you can get in this course? The lowest?
c. Compute the inverse of this equation. What would you need on the final to keep your final average above 80%?

Exercise 1.10 Let $f = ax + b$ be a linear function which is not a translation.

a. Show that f and f^{-1} have the same center; explain this geometrically.
b. If $f(x) = ax + (1 - a)c$ when written in slope-center form, show that its inverse, in slope-center form, is $f^{-1} = \frac{1}{a}x + (1 - \frac{1}{a})c$.

Exercise 1.11 Let $f(x) = ax + (1 - a)c$ and $g(x) = a'x + (1 - a')c$ be two linear functions with the same center c.

a. Put $g \circ f$ in slope-center form and observe that it also has c as its center.
b. Show that, if $f(x)$ and $g(x)$ are any two linear functions with the same center, then $f \circ g = g \circ f$.

Whenever $g \circ f = f \circ g$, we say that the functions f and g *commute*, or are *commutative*. As noted above, in general, one should not expect two functions to commute. On the other hand, you have just shown that when two linear functions have the same center, they do commute.

We close this section—and the chapter—by proving that this is the only case in which two non-translation linear functions commute.

Theorem 1.5 *Two linear functions commute if and only if one (or more) of the following cases hold:*

(1) *one of the functions is the identity;*
(2) *both of the functions are translations;*
(3) *both of the functions have the same center.*

Proof It follows from Theorem 1.4 (1) that two translations commute. In Exercise 1.11, you proved that two non-translations with the same center also commute. Additionally, one easily checks that the identity, $t_{[0]}$, commutes with every linear function. Thus, if any of the three cases hold, the linear functions commute.

Now assume that $f(x) = ax + b$ and $g(x) = a'x + b'$ are two linear functions which commute. We must show that one of the three cases hold. We do this by first computing both compositions:

$$(f \circ g)(x) = a(a'x + b') + b = aa'x + ab' + b,$$

$$(g \circ f)(x) = a'(ax + b) + b' = a'ax + a'b + b';$$

Setting them equal gives:

$$aa'x + ab' + b = a'ax + a'b + b'$$
$$ab' + b = a'b + b'$$
$$b - a'b = b' - ab'$$
$$(1 - a')b = (1 - a)b'$$

Suppose that $(1 - a') = 0$, i.e. $a' = 1$ so g is a translation. Then either $(1 - a) = 0$ or $b' = 0$. In the first instance, f is also a translation and case 2 holds; in the second instance, g is the identity and case 1 holds.

Similarly, if $(1 - a) = 0$, either case 1 or case 2 holds. We may assume then that both $(1 - a') \neq 0$ and $(1 - a) \neq 0$. Thus, neither f nor g are translations. Dividing through the last equality above by $(1 - a')(1 - a)$, we have:

$$\frac{b}{(1 - a)} = \frac{b'}{(1 - a')},$$

from which we conclude that f and g have the same center, i.e. case 3 holds.   □

Before we use these transformations of the line to develop the geometry of the Euclidean line, let us consider our traditional way of viewing linear functions as graphs, and interpret what we have done in this context.

To start with, the graph of the identity transformation $f(x) = x$ is the 45-degree line through the origin. The remaining lines with slope 1 are the translations of the identity, and their graphs are the lines parallel to the 45-degree line through the origin. In the following illustration, these lines are shown in green.

The graph of every other linear function intersects the 45-degree line in a unique point, and one easily checks that the coordinates of that point are (c, c), where c is the center of that linear function. Indeed, if $f(x) = ax + b$, then the intersection with the 45-degree line occurs where $f(x) = x$. Solving $ax + b = x$, we find $c = \frac{b}{1-a}$, so the intersection point is (c, c).

On the following grid, we have graphed:

- the 45-degree line (thick green),
- several translations (green),
- a reflection (blue),
- and two other similarities (red and orange).

It is always good to check your work with actual examples: the red graph corresponds to the function $f(x) = \frac{1}{3}x + \frac{4}{3}$. Solving $f(x) = x$, we find the center is $c = \frac{4/3}{1-1/3} = 2$, so the intersection point is $(2, 2)$, which matches the graph.

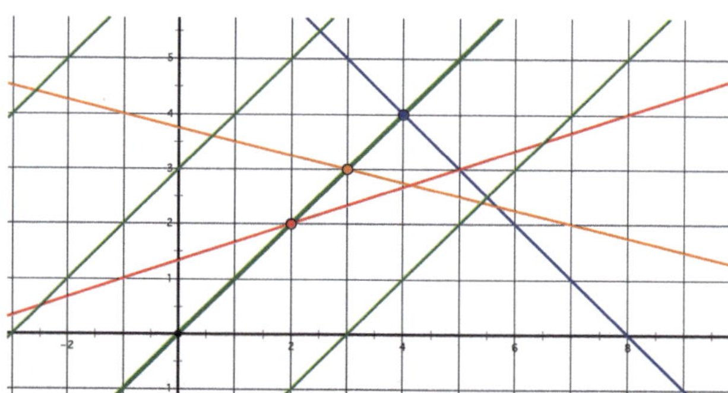

Exercise 1.12 Verify that the coordinates of the intersection of the red, blue and orange graphs are (c, c) where c is the center of the corresponding functions.

Exercise 1.13 Use the standard graph of the function $f(x) = ax + b$ to explain Theorem 1.3. Illustrate your explanation with the functions graphed above.

The Geometry of the Euclidean Line

2

2.1 The Euclidean Line

Since virtually all of our formal encounters with geometry have been with the Euclidean plane or Euclidean 3-space, we tend to think of the geometry of the Euclidean line as being trivial and uninteresting. In thinking this way, we are mistaken. While the Euclidean line is the simplest of the Euclidean spaces, its structure is far from trivial and is, in fact, rather interesting.

Among the simplest geometric objects on the Euclidean line—other than a single point—is the interval or segment. Intervals come in three "types":

- *Closed intervals*: two distinct points (numbers) and all points between them;
- *Open intervals*: all points between (but not including) two points;
- *Half-open intervals*: all points between two points including exactly one of the two endpoints.

The half-open intervals are further subdivided into those open on the right and those open on the left. We will adopt the convention that the endpoints of an interval are listed from smallest to largest, and we use the following notation:

- $[p, q]$ for the closed interval between p and q,
- (p, q) for the open interval,
- $[p, q)$ and $(p, q]$ for the two half-open intervals with endpoints p and q.

Our intuition tells us that two intervals of the same length and type should be congruent. For example, the intervals $[7, 12)$ and $[-7, -2)$ both have length 5 and are of the same type; therefore, they should be congruent.

J. J. Edmond, J. E. Graver, *The Linear Function and Euclidean Geometry*, Synthesis Lectures on Mathematics & Statistics, https://doi.org/10.1007/978-3-032-22712-6_2

To verify this using our definition of congruence, we must find a congruence of the line that maps $[7, 12)$ onto $[-7, -2)$. If we translate the line 14 units to the left, the point 7 is mapped to -7, the point 12 is mapped to -2, and all points between 7 and 12 are mapped to the points between -7 and -2. Thus, the translation

$$t_{[-14]}(x) = x - 14$$

maps $[7, 12)$ onto $[-7, -2)$, showing that the two intervals are congruent.

Next, consider the intervals $[7, 12)$ and $(-7, -2]$. Again, they are both half-open intervals of the same length. This time, the reflection

$$r_{\left[\frac{5}{2}\right]}(x) = -x + 5$$

through the point $\frac{5}{2}$ will work. Observe:

- 7, which is $4\frac{1}{2}$ units to the right of $\frac{5}{2}$, is reflected onto -2, which is $4\frac{1}{2}$ units to the left of $\frac{5}{2}$;
- 12, which is $9\frac{1}{2}$ units to the right of $\frac{5}{2}$, is reflected onto -7, which is $9\frac{1}{2}$ units to the left of $\frac{5}{2}$.

Moreover, all points between 7 and 12 are reflected onto the points between -7 and -2. Thus, this reflection maps $[7, 12)$ onto $(-7, -2]$, and the two intervals are congruent.

Exercise 2.1   Consider each of the 15 pairs of intervals from the following list of six intervals: $[-1, 3)$, $(4, 8)$, $[6, 10]$, $(-10, -6)$, $[8, 12]$, $(2, 6]$. In each case, decide if the two intervals are congruent. If they are congruent, find a congruence which maps one interval onto the other. Where possible, find a second congruence between the intervals.

Exercise 2.2   Let p, q and r be real numbers and assume that r is positive.

a. Describe the congruence which maps $[p, p + r)$ onto $[q, q + r)$.
b. Describe the congruence which maps $(p - r, p]$ onto $[q, q + r)$.
c. Describe both congruences which map $[p, p + r]$ onto $[q, q + r]$.

Exercise 2.3 Each interval is congruent to itself since it is mapped onto itself by the identity map $t_{[0]}$ which is a congruence. Which intervals can be mapped onto themselves by a congruence not equal to the identity? Describe the congruence.

The set of all numbers greater than or equal to (less than or equal to) a fixed number, is called a *closed ray* and the set of all numbers greater than (less than) a fixed number, is called an *open ray*. The notation we adopt is:

	Closed Rays	Open Rays

$$[p, \infty) = \{x \mid x \geq p\} \qquad (p, \infty) = \{x \mid x > p\}$$

$$(-\infty, p] = \{x \mid x \leq p\} \qquad (-\infty, p) = \{x \mid x < p\}$$

Exercise 2.4 Show that any two rays are congruent if they are either both open or both closed.

Now consider the two intervals $[7, 12)$ and $[21, 36)$. The second interval is three times as long as the first, which may lead us to suspect that they are similar. Specifically, we would like to find a similarity of the line that maps $[7, 12)$ onto $[21, 36)$.

Clearly, the magnification of this similarity must be 3. The most commonly known similarities are the *dilations about the origin* 0: each point x is mapped to ax, where a is the magnification factor of the dilation.

A dilation that shows these two intervals to be similar is the dilation with center 0 and magnification 3–that is x is mapped to $3x$. Under this dilation, 7 is mapped to 21, 12 to 36, and all points between 7 and 12 are mapped to the points between 21 and 36.

In general, finding the appropriate similarity that demonstrates two segments are similar is not always so straightforward. We will therefore defer consideration of more complicated examples until we have developed a deeper understanding of similarities.

2.2   Similarities

In Chap. 1, we proved that all linear functions are similarities of the Euclidean line. Although we did not prove it then, we stated that the converse is also true: all similarities of the Euclidean line are linear functions.

The fact that we could not think of any other similarities does not constitute a proof that none exist, and the proof itself is far from obvious. It involves several steps, the first of which is to show that a similarity is uniquely determined by its effect on any pair of distinct points.

We formalize this idea in the following key lemma.

Lemma 2.1 *Let f and g be two similarities and let p and q be two distinct points ($p \neq q$). If $f(p) = g(p)$ and $f(q) = g(q)$, then $f = g$.*

Proof We start the proof by demonstrating that f and g have the same magnification: Let a be the magnification of f. Let $p' = f(p) = g(p)$ and $q' = f(q) = g(q)$. Then

$$|g(p) - g(q)| = |p' - q'| = |f(p) - f(q)| = a|p - q|.$$

Thus, a is also the magnification of g. Next, we make the simple observation that $p' \neq q'$. Since $p \neq q$ and $a \neq 0$, we have: $|p' - q'| = |f(p) - f(q)| = a|p - q| \neq 0$, and we conclude that $p' \neq q'$.

To show that $f = g$, we must show that $f(z) = g(z)$ for all real numbers z. Suppose on the contrary that, for some z, $f(z) \neq g(z)$. We will show that this supposition leads to a contradiction and therefore cannot be true. Note that

$$|f(z) - p'| = |f(z) - f(p)| = a|z - p| = |g(z) - g(p)| = |g(z) - p'|,$$

i.e., p' is equally distant from $f(z)$ and $g(z)$. It follows that p' is the midpoint of the segment joining $f(z)$ and $g(z)$.

In short, we have shown that, if $f(z) \neq g(z)$, then p' is the midpoint of the segment with $f(z)$ and $g(z)$ as endpoints. Replacing p by q in this argument, we conclude that q' is also the midpoint of the segment joining $f(z)$ and $g(z)$. Thus, if $f(z) \neq g(z)$, then $p' = q'$ which we have already shown not to be true; contradiction! We conclude that, for all real numbers z, $f(z) = g(z)$ and that $f = g$. $\square$

Combining this lemma with Lemma 1.1 from the previous chapter, we can state and prove the main result of this section:

Theorem 2.1

(1) *A function f is a similarity of the real line if and only if f is a linear function.*
(2) *The absolute value of the slope of the linear function is the magnification of the similarity.*
(3) *A function f is a congruence of the real line if and only if f is a linear function with slope ± 1.*

Proof By Lemma 1.1, if f is a linear function, then it is a similarity.

Now assume that f is a similarity with magnification a; we must show that f is given by a linear function. If we can construct a linear function g which agrees with f at two distinct points, then, by Lemma 2.1, we can conclude that $f = g$, which shows that the similarity f is the linear function g.

Let $f(x) = ax + b$, then $b = f(0)$ and $a = f(1) - f(0)$. Since f is a similarity with magnification a, we have

$$|a| = |f(1) - f(0)| = a \cdot |1 - 0| = a,$$

so $a \neq 0$.

Define the linear function g by

$$g(x) = ax + b.$$

Then g and f are two similarities such that

$$g(0) = b = f(0)$$

and

$$g(1) = a(1) + b = a + b = \big(f(1) - f(0)\big) + f(0) = f(1).$$

By Lemma 2.1, it follows that $f = g$, and we conclude that f is a linear function. Furthermore, the magnification $a = |a|$ is the absolute value of the slope. Thus, f is a congruence if and only if $a = \pm 1$. $\qquad\square$

It is now clear from this theorem that translations and reflections are the only congruences of the Euclidean line. We state this fact explicitly in the following corollary.

Corollary 2.1 *The only congruences of the Euclidean line are the translations and the reflections.*

This theorem gives us a powerful algebraic tool for the study of similarities. For example, consider the problem discussed near the end of the last section: find a similarity which demonstrates that the intervals $[7, 12)$ and $[21, 36)$ are similar.

By the theorem, such a similarity must be given by a linear function. Suppose then that $f(x) = ax + b$ maps $[7, 12)$ onto $[21, 36)$. Clearly, 7 must map onto 21 and 12 must map onto 36. Thus, we have the system of equations

$$21 = f(7) = a \cdot 7 + b \quad \text{and} \quad 36 = f(12) = a \cdot 12 + b,$$

which is a system of two equations in the variables a and b.

Solving this system gives $a = 3$ and $b = 0$; so, as we had guessed,

$$f(x) = 3x$$

is the similarity.

Of course, we could have chosen to demonstrate the similarity by mapping $[21, 36)$ onto $[7, 12)$. In this case, we would compute the inverse of f:

$$f^{-1}(x) = \frac{1}{3}x.$$

Let's consider a more complicated example: find the similarities which map $[5, 6\frac{1}{2}]$ onto $[19, 23\frac{1}{2}]$. Since we may map 5 to 19 and $6\frac{1}{2}$ to $23\frac{1}{2}$ or 5 to $23\frac{1}{2}$ and $6\frac{1}{2}$ to 19, there are two possibilities. In both cases the interval $[5, 6\frac{1}{2}]$ of length $\frac{3}{2}$ is mapped onto the interval $[19, 23\frac{1}{2}]$ of length $\frac{9}{2}$, three times longer. We conclude that these similarity have slope 3 or -3. Which possibility corresponds with which slope has yet to be decided.

Suppose first that f has slope 3: $f(x) = 3x + b$. Then $f(5) = 15 + b$. To map 5 to 19, take $b = 4$; to map 5 to $23\frac{1}{2}$, take $b = 8\frac{1}{2}$. Checking we see that $f(x) = 3x + 4$ maps 5 to 19 and $6\frac{1}{2}$ to $23\frac{1}{2}$ and is a congruence of these intervals; one easily checks that while $f(x) = 3x + 8\frac{1}{2}$ maps 5 to $23\frac{1}{2}$, it maps $6\frac{1}{2}$ to 28 instead of 19 and is not a congruence of these intervals. Assuming the slope of g to be -3, $g(x) = -3x + b$, we again get two options for b: if $g(5) = 19$, $b = 34$; if $g(5) = 23\frac{1}{2}$, $b = 38\frac{1}{2}$. Again only one of these actually works. $g(x) = -3x + 38\frac{1}{2}$ maps 5 to $23\frac{1}{2}$ and $6\frac{1}{2}$ to 19.

Exercise 2.5 Give the one-line graphs for $f(x) = 3x + 4$ and $g(x) = -3x + 38\frac{1}{2}$, mark the intervals $[5, 6\frac{1}{2}]$ and $[19, 23\frac{1}{2}]$ and verify geometrically that f and g are congruences of these intervals.

Exercise 2.6 Consider each pair of intervals from the following list of intervals: $[-1, 5)$, $(4, 18)$, $[6, 10]$, $(-10, -3)$, $[8, 10]$, $(2, 6\frac{1}{2}]$. In each case, decide if the two intervals are similar. If they are similar, find a similarity between the two intervals. Where possible, find a second similarity between the intervals.

Exercise 2.7 Let p, q, r and s be real numbers with r and s positive.

a. Describe the similarity which maps $[p, p + r)$ onto $[q, q + s)$.
b. Describe the similarity which maps $[p, p + r)$ onto $(q - s, q]$.
c. Describe both similarities which map $[p, p + r]$ onto $[q, q + s]$.

2.3   Classification of Similarities

We next turn our attention to similarities of the Euclidean line which are not congruences. It follows from Lemma 1.4 that all similarities, except translations, have a fixed point. Drawing the one-line graph of such a function with the fixed point at the center motivates the following terminology.

If $a > 0$ and $a \neq 1$, we call the function $d(x) = ax + b$ a *dilation*, specifically, the dilation with center $\frac{b}{1-a}$ and magnification a. Using the slope-center form, we adopt the notation

$$d_{[a,c]}(x) = ax + (1 - a)c$$

for the dilation with center c and magnification a.

In this notation, for example, the linear function $f(x) = \frac{4}{5}x + 1$, which is the dilation with center 5 (since $5 = \frac{1}{1-\frac{4}{5}}$) and magnification $\frac{4}{5}$, is denoted by $d_{[\frac{4}{5},5]}(x)$.

The term "dilation" reflects the idea that the line is being expanded or contracted from the center c by a factor equal to the magnification.

Later in our investigations, we will wish to distinguish between those dilations with magnification greater than one and those with magnification between zero and one. We call the former *expansions* and the latter *contractions*.

Hence, $d_{[\frac{1}{2},-4]}$ is the contraction about -4 by the factor $\frac{1}{2}$, and $d_{[\frac{5}{2},\frac{7}{4}]}$ is the expansion about $\frac{7}{4}$ by the factor $\frac{5}{2}$.

How are we to interpret the similarity $f(x) = ax + (1-a)c$ when $a < 0$? Consider the example

$$f(x) = -\tfrac{8}{3}x - 7,$$

which, when rewritten in slope-center form, becomes

$$f(x) = -\tfrac{8}{3}x + \left(\tfrac{11}{3}\right)\left(-\tfrac{21}{11}\right).$$

This function not only stretches lengths by a factor of $\frac{8}{3}$, but also reflects points across the fixed point $-\frac{21}{11}$. Thus, it is the composition of the dilation $d_{[\frac{8}{3},-\frac{21}{11}]}$ and the reflection $r_{[-\frac{21}{11}]}$.

But in which order? Since both transformations have the same center, the order makes no difference. We proved this at the very end of Chap. 1 in Theorem 1.5. Hence,

$$f(x) = d_{[\frac{8}{3},-\frac{21}{11}]} \circ r_{[-\frac{21}{11}]} = r_{[-\frac{21}{11}]} \circ d_{[\frac{8}{3},-\frac{21}{11}]}.$$

Thus, when $a < 0$ and $a \neq -1$, we call

$$f(x) = ax + (1-a)c = d_{[|a|,c]} \circ r_{[c]}$$

the *dilating-reflection* with center c and magnification $|a| = -a$. We also extend the notation $d_{[a,c]}(x) = ax + (1-a)c$ to include the case when a is negative. (When $a = -1$, $d_{[a,c]}$ is simply the reflection $r_{[c]}$.)

The following corollary now follows directly from Theorem 2.1.

Corollary 2.2 *The only similarities of the Euclidean line which are not congruences are the dilations and the dilating-reflections.*

It follows from our classification of similarities that they naturally divide into those that preserve orientation and those that reverse it. Translations and dilations map half-open intervals that open to the right onto half-open intervals that also open to the right. In contrast, reflections and dilating-reflections map half-open intervals that open to the right onto half-open intervals that open to the left.

We say that translations and dilations are *direct* similarities, while reflections and dilating-reflections are *opposite* similarities.

In the first section of this chapter, we considered the problem of identifying the similarity which maps one interval onto another. Let us now reconsider this problem in light of what we have learned.

Suppose we wish to identify the similarity that maps $[5, 17)$ onto $[\pi, \pi + 12)$. Since the intervals have the same length, the magnification is 1; that is, the similarity is a congruence. Since the intervals also have the same orientation, the congruence is *direct*. Because translations are the only direct congruences, the similarity that maps $[5, 17)$ onto $[\pi, \pi + 12)$ must be a translation.

The only translation that maps 5 onto π is $t_{[\pi - 5]}$. Hence, $t_{[\pi - 5]}$ is the similarity that maps $[5, 17)$ onto $[\pi, \pi + 12)$.

This kind of reasoning will always work. Suppose we wish to identify the similarity f which maps $[5, 17)$ onto $(-2, 22]$. By comparing the lengths and orientations of the intervals, we conclude that f is an *opposite* similarity with magnification 2. Thus, f is a dilating-reflection and has the form $d_{[-2,c]}$. All that remains is to determine the center c.

We may do this in several ways; the simplest is to compute the value of $b = (1 - a)c$, where $a = -2$. Let $f = d_{[-2,c]}$. Then:

$$22 = f(5) = -2(5) + b \quad \text{so} \quad b = 32, \quad \text{and} \quad c = \frac{b}{1 - a} = \frac{32}{3}.$$

In general, given two points p and q, and a magnification $a > 0$, there are exactly two similarities with magnification a which map p onto q: one direct and one opposite. Assume $a \neq 1$, and let $d_{[a,c]}$, where $|a| = \pm a$, denote such a similarity. Then:

$$q = d_{[a,c]}(p) = ap + (1 - a)c \quad \text{so that} \quad q - ap = (1 - a)c.$$

Thus, the center c may be computed directly from a, p, and q using the formula:

$$c = \frac{q - ap}{1 - a}.$$

We now formalize this for later reference:

Lemma 2.2 *Given real numbers p, q and a positive real number m.*

(1) *If $a \neq 1$ then the two similarities which map p onto q are the dilation $d_{[a, \frac{q-ap}{1-a}]}$ and the dilating-reflection $s_{[-a, \frac{q+ap}{1+a}]}$.*

(2) *If $a = 1$ then the two similarities which map p onto q are the translation $t_{[q-p]}$ and the reflection $r_{[\frac{q+p}{2}]}$.*

Exercise 2.8 Find both similarities which map $[\sqrt{2}, \sqrt{72}]$ onto $[\sqrt{2} - 5, \sqrt{2} + 5]$.

Exercise 2.9 Let p and q be two distinct real numbers and let a be a positive real number.

a. Find the set of points x satisfying $d_{[x,q]} = a d_{[x,p]}$.
b. Explain how the points in this set are related to the similarities that map p onto q.

Exercise 2.10

a. Draw the traditional graph for each of several different linear functions. Include a translation a reflection and at least one dilation.
b. Use the properties of the traditional graph to explain why, translations have no fixed points while all other linear functions have a unique fixed point.

2.4   The Algebra of Transformations

Given two similarities of the line, it is natural to consider their composition. For example: what is the result of reflecting about 0 and then reflecting about 3? More generally, how do we identify the composition of two congruences or similarities?

By combining Theorem 2.1 with Lemma 1.3, we know that the composition of two similarities is itself a similarity. Additionally, by combining Theorem 2.1 with Theorem 1.1, we see that the inverse of a similarity is also a similarity.

There are several further observations that can be made. For instance, the composition of two congruences is always another congruence.

Exercise 2.11 Explain in words why:

a. the composition of two direct similarities is a direct similarity;
b. the composition of two opposite similarities is a direct similarity;
c. the composition of one direct and one opposite similarity, in either order, is an opposite similarity.

Since by now the reader should excel at all the necessary tools to investigate similarities of the Euclidean line, we leave many of the proofs of the results in this section as exercises for the reader. We begin by considering the algebra of congruences.

Theorem 2.2 *Consider the translations $t_{[b]}(x)$ and $t_{[b']}(x)$ and the reflections $r_{[c]}(x)$ and $r_{[c']}(x)$. Then*

(1) The inverse of $t_{[b]}$ is the translation $t_{[-b]}$.

(2) The inverse of $r_{[c]}$ is the reflection $r_{[c]}$ itself.

(3) The composition of $t_{[b]}$ and $t_{[b']}$, in either order, is the translation by $(b + b')$: $t_{[b+b']}$.

(4) The composition $r_{[c]} \circ r_{[c']}$ is the translation by $2(c - c')$: $t_{[2(c-c')]}$.

(5) The composition $t_{[b]} \circ r_{[c]}$ is the reflection with center $c + \frac{b}{2}$: $r_{[c+\frac{b}{2}]}$.

(6) The composition $r_{[c]} \circ t_{[b]}$ is the reflection with center $c - \frac{b}{2}$: $r_{[c-\frac{b}{2}]}$.

Note that you have already proved several parts of this theorem, but we included the statements here for the sake of completeness.

Exercise 2.12 Prove the remaining parts of Theorem 2.2.

Corollary 2.1 *Any isometry can be written as the composition of two or fewer reflections*

Proof We say that a reflection is the "composition" of that one reflection. By 4) above, the translation $t_{[b]} = r_{[\frac{b}{2}]} \circ r_{[0]}$, in particular the identity map is $r_{[c]} \circ r_{[c]}$, for any c. $\square$

The "algebra" of arbitrary similarities is described in the following theorem.

Theorem 2.3

(1) $d_{[a,c]} \circ d_{[a',c]} = d_{[aa',c]}$, *when $aa' \neq 1$.*

(2) $d_{[a,c]} \circ d_{[\frac{1}{a},c]} = t_{[0]}$

(3) $d_{[a,c]} \circ d_{[a',c']} = d_{[aa',c'']}$ *for some point c'' whenever $aa' \neq 1$.*

(4) $d_{[a,c]} \circ d_{[a',c']} = t_{[b]}$ *for some point b whenever $aa' = 1$.*

Exercise 2.13 Prove this theorem.

Exercise 2.14 Compute a formula for c'' in part 3 and b in part 4 of this theorem.

The set of all congruences of the line is *closed* under composition and taking inverses: the composition of two congruences is a congruence, and the inverse of a congruence is also a congruence. Any set of functions satisfying these two conditions is called a *group*.

We will give a formal definition of a group in Part II; but for the purposes of this section, these two criteria suffice to define the group. Thus, the set of all congruences is called the *group of congruences*.

There are many subsets of the group of congruences which are themselves groups. To verify that a nonempty subset is a group, it suffices to check that the inverse of any element in the set is also in the set, and that the composition of any two elements in the set is again in the set.

It follows from the theorem that the set of all translations is such a group, often called a *subgroup*, since it is a group contained within another group. Note that the set of reflections is not a subgroup: it is closed under inverses but not under composition.

Exercise 2.15 Which of the following collections of similarities are subgroups?

a. The collection of all direct similarities.
b. The collection of all opposite similarities.
c. The collection of all similarities with the same center c.
d. The collection of all similarities with integral magnifications.
e. The collection of all similarities with integral centers.
f. The collection of all translations by an integral amount.

One of the sources of interesting subgroups are the subsets of the line: given a subset S, one may consider the collection of all congruences (or all similarities) which map S onto itself. Clearly, the composition of two congruences (or similarities) in this collection is also in the collection, and the inverse of any congruence (or similarity) in the collection is also in the collection.

The collection of all similarities which map S onto itself is called the *symmetry group of S*. We denote the symmetry group of S by $\mathcal{G}(S)$.

For most sets S, $\mathcal{G}(S)$ contains just one symmetry: the identity $t_{[0]}$. We say that such a set is *asymmetric*. A half-open interval is asymmetric. However, in an earlier exercise you showed that a closed or an open interval is not asymmetric; although the symmetry group of such an interval is quite simple, consisting of just two congruences—the identity and the reflection about the midpoint of the interval.

Exercise 2.16 Describe the symmetry groups of the following sets.

a. $\mathbb{Z}$, the set of all integers.
b. P, the set of all positive integers.
c. O, the set of all odd integers.
d. I, the set of all intervals of the form $[p, p + 1)$ where $p \in O$.
e. T^{+}, the set of all integer powers of 2, i.e. all numbers of the form 2^z where $z \in Z$.
f. $R = [r, \infty)$.
g. T, the set of all positive and negative powers of 2, i.e. all numbers of the form $\pm 2^z$ where $z \in Z$.
h. $\mathbb{Q}$, the set of all rational numbers.
i. $U = [r, \infty) \cup (\infty, s]$ where $s < r$.

Linear Difference Equations 3

3.1 Difference Equations

In this and the next chapter, we will explore applications of linear functions to personal finance, including savings, annuities, and loans. To motivate our definitions, we begin with an example.

Suppose you purchase a new car and take out a four-year, \$30,000 loan. If the monthly interest rate is 1% (equivalent to a 12% annual rate), your monthly payments will be \$790.02. (We will develop the formula used to calculate loan payments in the next chapter.)

Now suppose that after two years, you "come into some money" and wish to pay off the loan early. How much do you still owe? More generally, how can we determine the remaining balance on the loan at any given time?

The balance due on this loan can be thought of as a sequence of numbers: $b_0, b_1, \cdots, b_{48}$, where b_0 is the amount of the loan and b_i is the balance due at the end of the ith month, for $i = 1, \cdots, 48$. The problem then is to compute the numbers in this sequence.

As it turns out, each term is easily computed from the previous term. As noted above $b_0 = 30,000$. At the end of the first month, you owe the \$30,000 plus the accrued interest of \$300 ($.01 \times 30,000$) minus the first payment of \$790.02, or \$29,509.98, ($30,000 + 300 - 790.02 = 29,509.98$). The amount due at the end of the second month is \$29,209.98 plus the accrued interest \$295.10 ($.01 \times 29,209.98$) minus the second payment or \$28,715.06. In general, b_{n+1}, the balance due at the end of the n plus first month, is calculated as b_n, the balance due at the end of the previous month plus $.01 \times b_n$, the accrued interest, minus \$263.66, the next payment. Since $b_n + (.01)b_n = (1.01)b_n$, the balance due on the loan is given by the equations:

$$b_{n+1} = (1.01)b_n - 790.02, \quad \text{for } n = 1, \cdots 48.$$

© The Author(s), under exclusive license to Springer Nature Switzerland AG 2026   29
J. J. Edmond, J. E. Graver, *The Linear Function and Euclidean Geometry*, Synthesis
Lectures on Mathematics & Statistics, https://doi.org/10.1007/978-3-032-22712-6_3

As we have seen, the balance due decreases each month and, if the payments are the correct size, b_{48} will be zero. The balance due on a loan can be placed in a more general context: Consider a linear function $f(x) = ax + b$ and an initial value x_0. Then study the *trajectory* of x_0 under repeated applications of f, that is, the sequence of points

$$x_0$$

$$x_1 = f(x_0)$$

$$x_2 = f(x_1) = f^2(x_0)$$

$$\vdots$$

$$x_n = f(x_{n-1}) = f^n(x_0)$$

$$\vdots$$

This way of looking at linear equations has many applications and its own special terminology. The sequence $\{x_i \colon i = 0, 1, \ldots\}$ described here is called the *solution* to the *linear difference equation* $x_{n+1} = ax_n + b$ with *initial value* x_0.

There is a linear difference equation associated with every linear function. Let's consider, from this perspective, some specific linear functions. First, consider the translation $t_{[1]}$. If we take x_0 to be 0, then $x_1 = 1$, $x_2 = 2$, and one easily sees that $x_n = n$ for all positive integers n. In fact, the difference equations associated with translations always have such simple solutions. Consider the linear difference equation corresponding to $t_{[b]}$, i.e. $x_n = t_{[b]}(x_{n-1}) = x_{n-1} + b$. We immediately find:

$$x_1 = x_0 + b,$$

$$x_2 = x_1 + b = x_0 + 2b,$$

$$x_3 = x_2 + b = x_0 + 3b, \, etc.$$

It is clear that the solution is given by the formula $x_n = x_0 + nb$, for all positive integers n. Hence, the solution to a linear difference equation which corresponds to the translation $t_{[b]}$ and the initial value x_0 is an *arithmetic progression* starting at x_0 with difference b. Conversely, every arithmetic progression is the solution to a linear difference equation corresponding to a translation.

If the solution sets to linear difference equations associated with translations are simple, the solution sets to linear difference equations associated with reflections are even simpler, as you will see upon working the next exercise.

Exercise 3.1

a. Solve the linear difference equation associated with the reflection $r_{[4]}$ for each of the following initial values:

$$x_0 = 0, \quad x_0 = -7, \quad x_0 = \frac{19}{2}, \quad x_0 = \sqrt{3}.$$

b. Let $r_{[c]}$ be the reflection in the point c, so that the associated linear difference equation is

$$x_n = -x_{n-1} + 2c.$$

Show that the solution to this equation, for any initial value x_0, is given by:

$$x_n = \begin{cases} x_0, & \text{if } n \text{ is even,} \\ 2c - x_0, & \text{if } n \text{ is odd.} \end{cases}$$

Thus, when the associated linear function is a *congruence*, the solutions to a linear difference equation are quite easy to describe. However, these solutions behave very differently from those arising from *similarities* that are not congruences. For this reason, such cases are typically excluded from the standard definition of a linear difference equation.

From this point forward, we will adopt the common convention of considering only linear difference equations of the form

$$x_{n+1} = ax_n + b,$$

where $a \neq 0, 1,$ or -1. In other words, we will exclude translations, reflections, and constant functions.

With these exclusions, linear difference equations fall into four natural categories, depending on the value of a: $a \in (-\infty, -1)$, $a \in (-1, 0)$, $a \in (0, 1)$, or $a \in (1, \infty)$.

As we will soon see, the behavior of the solutions within each category exhibits distinctive and consistent patterns, making this classification both natural and useful.

Another interesting example of a linear difference equation comes from the physics of a bouncing ball. Consider the following experiment: drop a ball from a fixed height and measure the height of its first bounce. For consistency, measure the distance from the floor (or tabletop) to the bottom of the ball. Repeating this simple experiment several times at several different heights, one typically finds that the height of the rebound from an initial height x is given by a linear function of the form $r(x) = ax$, where a is a constant satisfying $0 < a < 1$.

Once we have determined a experimentally, we can use it to predict the behavior of the ball as it continues to bounce. Suppose the ball is dropped from a height x_0. It will rebound to a height of $x_1 = ax_0$. The second rebound will reach a height of $x_2 = ax_1 = a^2x_0$. In general, the height of the nth rebound, x_n, satisfies the linear difference equation

$$x_n = ax_{n-1}.$$

In this case, the solution is given explicitly by the formula

$$x_n = a^n x_0.$$

You can experimentally verify this prediction by taping a sheet of paper to a wall, marking the initial height x_0 and the predicted heights $x_1, x_2, \ldots$, then dropping the ball from x_0 and observing the actual rebound heights.

This example illustrates a simple case in which the linear difference equation has the form $x_{n+1} = ax_n$, with $b = 0$, making the formula for x_n easy to compute. However, when $b \neq 0$, deriving the general formula for x_n becomes more involved. This will be the focus of the next section.

3.2 Formula for the nth Term

Return to the loan problem with which we opened this chapter. Recall that what we really wanted to determine was the amount due on the loan after two years—or equivalently, after 24 months—which we denoted by b_{24}. It is natural to ask whether we must compute all the previous terms $b_1, b_2, \ldots, b_{23}$ in order to find b_{24}. The answer is no.

The main goal of this section is to develop a formula for the nth term in the solution to a linear difference equation. Since these terms are defined inductively, one approach to constructing such a formula is to express each term as a sum, as illustrated below, and then apply the formula for the sum of a geometric progression to derive the required expression.

$$\begin{aligned}
x_1 &= ax_0 + b \\
x_2 &= a^2x_0 + ab + b \\
x_3 &= a^3x_0 + a^2b + ab + b \\
&\ \ \vdots \\
x_n &= a^nx_0 + a^{n-1}b + \cdots + ab + b
\end{aligned}$$

Another way to present the formula is to simply state it and then prove by induction that it is the correct expression for the solution. In general, this method is the most efficient. However, it is often not very satisfying, since arriving at the formula in the first place seems to require already knowing the answer.

The first approach—deriving the formula as a sum and using the formula for a geometric progression—is certainly more satisfactory, and it serves as a nice application of geometric series. However, even this method produces the formula without fully explaining why it should be expected.

Our geometric approach to linear functions, on the other hand, will not only yield the correct formula, but also clearly reveal why it must hold.

First, let us consider two of the simpler cases that we have previously set aside.

The translation $t_{[b]}$ moves each point b units to the right if $b > 0$, or $|b|$ units to the left if $b < 0$. Thus, one immediately sees that the terms of the solution sequence are equally spaced, $|b|$ units apart, and are given by:

$$x_0, \; x_0 + b, \; x_0 + 2b, \; \ldots$$

This is an arithmetic progression.

In the case of the reflection $r_{[c]}$, one similarly observes that the images of a point x_0 under successive applications of $r_{[c]}$ alternate between x_0 and its reflection about c, which is $2c - x_0$.

Now, let us consider the linear difference equation associated with the dilation $d_{[a,c]}$. Below, we illustrate two specific cases. In the first case, we take:

$$a = \tfrac{3}{2}: x_n = \tfrac{3}{2}x_{n-1} + 1; \; x_0 = 0.$$

In the second case, we take:

$$a = \tfrac{3}{4}: x_n = \tfrac{3}{4}x_{n-1} + 1; \; x_0 = 12.$$

Consider the first of these dilations, where the magnification factor is $\tfrac{3}{2}$. Since $c = -2$ is the fixed point and distances from c are multiplied by $\tfrac{3}{2}$ at each step, the distance between x_1 and c is $\tfrac{3}{2}$ times the distance between x_0 and c. Similarly, the distance between x_2 and c is $\tfrac{3}{2}$ times the distance between x_1 and c, or $(\tfrac{3}{2})^2$ times the distance between x_0 and c.

By induction, we see that the distance between x_k and c is

$$\left(\frac{3}{2}\right)^k \cdot |x_0 - c|.$$

The same argument applies to the second case, where the magnification is $\tfrac{3}{4}$. In this case, the distance between x_k and c is

$$\left(\frac{3}{4}\right)^{k} \cdot |x_0 - c|.$$

These two examples illustrate the behavior of the sequence of points generated by successive applications of the dilation $d_{[a,c]}$. In the first case, where $a = \frac{3}{2} > 1$, each term in the sequence moves farther away from the fixed point c. If the initial point x_0 were less than c, the terms would still move away from c, but in the negative direction.

In the second example, where $a = \frac{3}{4} < 1$, each term moves closer to c, and in fact, the sequence converges to c. Again, if x_0 were less than c, the terms would approach c from below.

Now consider the linear difference equation

$$x_n = ax_{n-1} + b,$$

where $a \neq 0, \pm 1$, associated with the linear function $f(x) = ax + b$.

Write f in slope-center form:

$$f(x) = ax + (1 - a)c,$$

and rewrite the difference equation in this form:

$$x_n = ax_{n-1} + (1 - a)c.$$

This gives:

$$x_n - c = a(x_{n-1} - c), \quad \text{for all } n.$$

Thus, we have:

$$x_1 - c = a(x_0 - c),$$

$$x_2 - c = a(x_1 - c) = a^2(x_0 - c),$$

$$x_3 - c = a(x_2 - c) = a\big(a^2(x_0 - c)\big) = a^3(x_0 - c),$$

$$\vdots$$

$$x_n - c = a(x_{n-1} - c) = a\big(a^{n-1}(x_0 - c)\big) = a^n(x_0 - c).$$

This is formalized in the following theorem.

Theorem 3.1 *Let $x_0, x_1, \ldots, x_n, \ldots$ be a solution to the linear difference equation $x_n = ax_{n-1} + b$ (where $a \neq -1, 0, 1$); then $x_k = a^k(x_0 - c) + c$, where $c = \frac{b}{1-a}$.*

Corollary 3.1 *Let $x_0, x_1, \ldots, x_n, \ldots$ be a solution to the linear difference equation $x_n = ax_{n-1} + b$ (where $a \neq -1, 0, 1$).*

(1) *If $0 < a < 1$, then the solution is a monotone sequence of points starting at x_0 and converging toward $\frac{b}{1-a}$.*

(2) *If $1 < a$, then the solution is a monotone sequence of points starting at x_0 and diverging from $\frac{b}{1-a}$.*

(3) *If $-1 < a < 0$, then the solution is a sequence of points starting at x_0 and converging toward $\frac{b}{1-a}$, but alternating from one side of $\frac{b}{1-a}$ to the other.*

(4) *If $a < -1$, then the solution is a sequence of points starting at x_0 and diverging from $\frac{b}{1-a}$, but alternating about $\frac{b}{1-a}$.*

Exercise 3.2 For each of the following linear difference equations with each of the given initial values, (1) find the fixed point c, (2) write out the formula for x_n, (3) describe the solution sequence as monotone or alternating and converging or diverging, (4) graph the first several terms of the solution.

a. $x_n = \frac{7}{5}x_{n-1} - 1$ for $x_0 = 0, 5, 10$

b. $x_n = -\frac{7}{5}x_{n-1}1$ for $x_0 = 0, \frac{1}{2}, 2$

c. $x_n = \frac{5}{7}x_{n-1} - 1$ for $x_0 = 0, 5, 10$

d. $x_n = -\frac{5}{7}x_{n-1} - 1$ for $x_0 = 0, \frac{1}{2}, 2$

Exercise 3.3 Now that you know the formula for the nth term of the solution to a linear difference equation, give an induction proof for Theorem 3.1.

3.3 Some Simple Applications

3.3.1 Population Models

Consider the following problem: you own a piece of land containing a pond which you would like to stock with trout. Trout will not reproduce in this pond but roughly two-thirds of the fish population will survive from one year to the next. Suppose that you initially stock the pond with 5000 fish and then add 2000 in each successive year. At what level will the fish population stabilize? One easily sees that the population levels in your pond are given by a linear difference equation: if p_n denotes the size of the fish population in the nth year, just after the new fish have been added, then the sequence p_n is the solution to the linear difference equation $p_n = \frac{2}{3}p_{n-1} + 2000$ with initial condition $p_0 = 5000$. The fixed point for this difference equation is $c = \frac{2000}{1-\frac{2}{3}} = 6000$. Thus, in a few years, the fish population should stabilize at about 6000 fish. Note that, contrary to our intuition, this stable value does not depend on the size of the initial stocking. The size of the initial stocking only affects how long it takes the population to approach the stable value. These observations are illustrated in the following exercises.

Exercise 3.4 All of the fish that you have been using to stock your pond come from a friend who owns a pond in which trout do reproduce. Each year the trout population in his pond increases by 25% until the maximum population of 12,000 trout is reached. When you initially stocked your pond with 5000 trout from his pond, his was at its capacity of 12,000. If we let q_n denote the number of trout in your neighbor's pond, we have then that $q_0 = 7000$.

a. Give the linear difference equation that governs the population size in your friend's pond.
b. Assume that you continue to stock your pond with 2000 trout each year from your friend's pond. What will the trout population be in your friend's pond after 10 years?
c. How could this have been avoided?

Exercise 3.5 Again consider the pond described at the beginning. We say that the population has stabilized once it is within 100 of the fixed value.

a. In the initial problem, how many years will it take for the fish population to stabilize?
b. If the pond is initially stocked with 10,000 fish, how many years will it take for the population to stabilize?
c. What should be the size of the initial stocking if you wish the population to stabilize as soon as possible? Formulate your answer as a general principle.
d. If you wish the stable population in your pond to be 10,000, how many fish should be added each year?

There are innumerable variations on this "population model". The population to be modeled could be fish or some other animals or people. In such models, the subscript n denotes the number of time periods: for people and larger animals n often stands for the number of years or decades which have elapsed, for smaller animals (mice, for example) n could stand for the number of months, days may be appropriate for insects such as roaches and hours for bacteria. The magnification a is usually greater than 1 and $100 \times (a - 1)$ represents the percentage population increase (births minus deaths divided by population) for the given time period due to the natural biological processes. The constant b then represents the change in population due to immigration and emigration (immigration minus emigration, to be precise). Whenever $b \geq 0$ successive populations x_n will grow unbounded indicating that the population will eventually out grow its environment. At this point, our simple linear difference equation model becomes invalid. For such models, the relevant question is: how long will it take the population to reach the environmental limits? Another question often asked of such models is: how long will it take for the population to double? In computing the formula for the nth term of a solution to a difference equation, exponential functions arise (in general, unconstrained population growth is exponential). It is not surprising then that, in solving the questions just posed, logarithmic functions will arise.

Exercise 3.6 (Research Project) Find population records for the United States beginning in the year 2000. Using this data, determine constants a and b to construct a population model of the form

$$p_n = ap_{n-1} + b.$$

Assume p_0 corresponds to the population in the year 2000. Let n denote the number of lapsed decades. Then p_n is the population in $20n0$ for $n = 0, \cdots, 9$, and predicted populations for the years $2000 + 10n$ when $n \geq 10$. As stated above, $100 \times (a - 1)$ represents the percentage population increase (births minus deaths) for the given time period due to the *natural biological processes* - be careful here, your data may include the immigration/emigration figures. Once you have constructed your model answer the following questions.

a. When does your model predict that the population of the US will reach 300,000,000?
b. How long will it take for the year 2000 population to double?

Exercise 3.7 Consider the population model $p_n = ap_{n-1} + b$

a. Find the formula for the number of time periods needed for the population to double assuming that $b = 0$.
b. Find the formula for the number of time periods needed for the population to double assuming that $b \neq 0$.

3.3.2   Radioactive Decay

Certain physical processes may also be described by linear difference equations. One classic example is radioactive decay. Given a sample of a radioactive substance, its level of radioactivity decreases by a fixed percentage each year. For example, assume that the level of radioactivity for this substance decreases by one half of one percent each year. Thus, from one year to the next, it retains 99.5% of its radioactivity and the governing linear difference equation is $r_n = .995r_{n-1}$ where r_i is the measure of the radioactivity of the sample of the substance after i years. One question that we may ask is: how long will it take for the radioactivity of this sample to drop to one half of its original value? Using our formula for the nth term, we conclude that $r_n = (0.995)^n r_0$ and we see that r_n will equal $\frac{1}{2} \times r_0$ when $(0.995)^n = 0.5$. Taking the natural log of both sides, we get:

$ln(0.995^n) = n\ ln(0.995) = ln(0.5)$, and solving for n: $n = \frac{ln(0.5)}{ln(0.995)} = 138.28$. We notice that the 138.28 years does not depend on the size of the sample, the initial radiation level r_0 or by what units radiation is measured. This number, called the *half-life* of the substance and is a constant based on the properties of the substance itself.

Exercise 3.8

a. If a given radioactive substance has a half-life of 200 years, by what percentage will its radioactivity be reduced each year?
b. Look up the half-life for several radioactive substances and write out the difference equations for their radioactivity levels over time.

3.3.3   The Towers of Hanoi

For an entirely different application, we consider the mathematical puzzle called the Towers of Hanoi. The puzzle consists of three posts and a set of n disks which have holes in their centers and slide over the posts. The disks are all of different diameters and are arranged from the largest (on the bottom) to the smallest (on the top) on post A. (See the diagram below.)

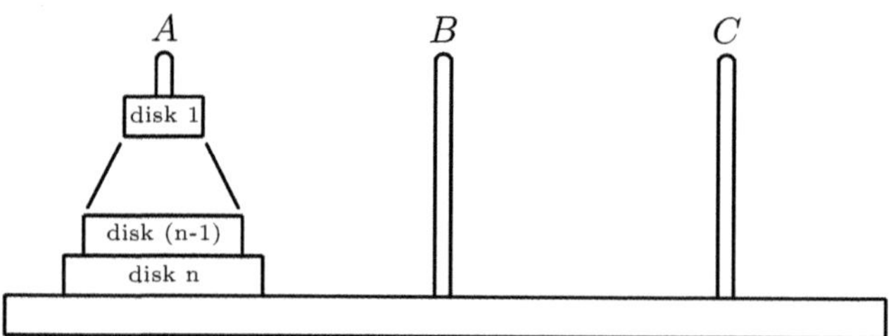

The object of the puzzle is to transfer all of the disks from post A onto post B in as few moves as possible. Moves are governed by the following rules:

1. Only one disk may be moved at a time;
2. All disks not being moved must be on one of the posts;
3. No disk may be placed on top of a smaller disk.

If the number of disks is 1, we simply move disk 1 from post A to post B in 1 move. It is not too hard to see that you actually need 3 moves to transfer 2 disks from post A to post B. We "park" disk 1 on post C (1 move), then move disk 2 to post B and finally move disk 1 to post B for a total of 3 moves. Let m_i denote the least number of moves needed to transfer i disks from one post to another; so far we have $m_1 = 1$, $m_2 = 3$ and, of course, $m_0 = 0$. Now, in order to move disk n from post A to post B, we must park disk 1 through disk $(n-1)$ on post C. This can be done most efficiently in m_{n-1} moves. Disk n is then move to post B. Finally, the the disks on C are transferred to post B, again in m_{n-1} moves. The total number of moves needed to transfer n disks from A to B is $m_{n-1} + 1 + m_{n-1}$ and m_n is given by the difference equation: $m_n = 2m_{n-1} + 1$.

Exercise 3.9 Solve this linear difference equation and give the formula for m_n. Verify that the formula works for $n = 0, 1, 2, 3$ and 4.

The story behind the name "Towers of Hanoi" is that the transfer of 64 gold disks is actually being carried out in a Buddhist monastery near Hanoi. The monks started at the beginning of the world and the world will end when the transfer is completed.

Exercise 3.10 There is a ceremony for the moving of a disk so each single move requires exactly one minute. Furthermore, according to the story, the world began exactly 12,544 years ago.

a. Should we be worried about the world ending in our lifetime?
b. How long would a "24-disk world" last?
c. How many disks would a 1 billion year world need?

Things could be sped up (although it is not clear that we would want to do this) if there was another parking post, 4 posts in all. Finding the best strategy in the four post problem can be challenging. The solution to this is not of direct interest to study of the linear function, however the following question is: let f_n denote the least number of moves needed to transfer n disks from one post to another post when there are a total of 4 posts available, is the sequence f_n the solution set of a linear difference equation?

Exercise 3.11 Compute directly f_1, f_2, f_3 and f_4 and see if you can find a linear difference equation which will give these values. If you cannot, prove that it cannot be done; if you can, compute f_5 directly and from the difference equation and compare the results.

Given any sequence of real numbers $x_0, x_1, \cdots, x_n, \cdots$, one may ask if it is the solution set of a linear difference equation.

Exercise 3.12 Prove that $x_0, x_1, \cdots, x_n, \cdots$ is the solution set of a linear difference equation if and only if, for each integer $i \geq 0$, we have: $(x_{i+3} - x_{i+2})(x_{i+1} - x_i) = (x_{i+2} - x_{i+1})^2$.

3.3.4   Weight Management

An interesting application of linear difference equations is weight management. Let's consider Jack, who currently weighs 176 pounds. Due to concerns about high blood pressure, his doctor has recommended that he aim to reduce his weight by 10%, targeting 158 pounds. While reducing calorie intake can lead to short-term weight loss, Jack's doctor knows it doesn't always result in sustainable, long-term success. Over time, the body may adapt by slowing down metabolism, making it harder to continue losing weight or maintain the loss. A balanced approach, including both dietary changes and regular physical activity, is often more effective for long-term weight management. Nevertheless, for simplicity of this example, Jack and his doctor decide that Jack should try to achieve his target weight by reducing his calories.

Jack maintains his current weight of 176 pounds by consuming an average of 2480 calories per day. Dividing 2480 by 176, we see that Jack needs approximately 14.1 calories per pound to maintain his weight. While we know that weight loss is not linearly effected by calorie intake, let's assume this ratio remains constant as long as Jack maintains the same diet quality and activity level.

Jack's initial reaction was: "This isn't too difficult; all I need to do is reduce my daily intake by 10% to 2232 calories per day." The doctor acknowledged that Jack's calculations were correct but pointed out that, while a diet of 2232 calories per day would maintain a weight of 158 pounds, it would take a considerable amount of time to reach that weight on such a diet. Jack hadn't considered another crucial factor: the number of calories he needs to "burn off" to lose 1 pound of weight. Based on Jack's lifestyle and metabolism, the doctor estimated that it would take a cumulative reduction of 3500 calories to lose 1 pound.

We know that a too severe reduction of calories is unhealthy and has not been shown to lead to permanent weight loss. So Jack and the doctor agreed that a safe goal would be to reach 158 pounds over the next 3 months (90 days). Jack, who is good with arithmetic, calculated $\frac{18 \times 3500}{90} = 700$ and $2480 - 700 = 1780$, concluding that if he reduced his daily intake to 1780 calories per day, he would reach his target weight in 90 days. However, the doctor informed Jack that he would need to follow a 1650-calorie per day diet instead. Jack was surprised and asked why his calculations were incorrect.

Exercise 3.13 Explain the error in Jack's reasoning. If needed, there is a hint at the end of this section.

Once Jack understood the error in his logic, he asked the doctor where the 1650-calorie figure came from. The doctor showed Jack a table and explained that the entries were computed using a linear difference equation. Jack went home, learned about difference equations, and developed the following (correct) analysis.

Let d denote Jack's daily caloric intake, 2480 now, and some other value in the future yet to be computed. Let x_0 denote Jack's weight today ($x_0 = 176$ lbs) and let x_n denote

Jack's weight n days from now. If all goes well, x_{90} will equal 158 lbs. For the nth day, we make the following computation:

$$\frac{d - 14.1x_{n-1}}{3500} + x_{n-1} = x_n.$$

Here, $d - 14.1x_{n-1}$ is the calorie surplus (when positive) or deficit (when negative) on the nth day, and this results in a $\frac{d-14.1x_{n-1}}{3500}$ pound weight change (gain when positive, loss when negative) on the nth day. Thus, Jack's daily weight is given by the linear difference equation:

$$x_n = \frac{3485.9}{3500}x_{n-1} + \frac{d}{3500}.$$

The fixed point of this equation is easily computed to be $\frac{d}{14.1}$. So, if $d = 2232$ (14.1×158, Jack's initial plan), Jack's weight will eventually approach 158 lbs. However, since the slope $\frac{3485.9}{3500} \approx 0.9959$ is so close to 1, it would indeed take years to get within a half-pound of 158 lbs.

Exercise 3.14 If Jack were to go on a 2232-calorie diet, what would he expect his weight to be in 1 year? In 2 years? In 3 years?

Exercise 3.15 Set up the difference equation for a diet of 1860 calories per day and track Jack's weight at 30-day intervals until it reaches 158 lbs.

Returning to the difference equation with d unspecified, we compute the value of the nth term:

$$x_n = \left(\frac{3485.9}{3500}\right)^n \left(x_0 - \frac{d}{14.1}\right) + \frac{d}{14.1}.$$

Substituting 176 for x_0 and solving for d gives:

$$d = \frac{14.1\left(x_n - \left(\frac{3485.9}{3500}\right)^n 176\right)}{1 - \left(\frac{3485.9}{3500}\right)^n}.$$

Now substituting 158 for x_n and 90 for n, we compute d to be 1651, which the doctor rounded to 1650 in the table.

Exercise 3.16 Crash diets can are designed to achieve rapid weight loss, but they are unsustainable and may pose serious health risks, including nutrient deficiencies, muscle loss, and a slowed metabolism. These often involve dangerous reductions in caloric intake.

How many calories per day would Jack be allowed if he wanted to achieve his 10% weight reduction in 60 days? What about 30 days? Do these seem like safe amounts?

Exercise 3.17 For most individuals, there are two key constants, denoted c and C: c is the number of calories needed to maintain 1 pound of a person's weight, and C is the number of calories required to change a person's weight by 1 pound. For Jack, these are 14.1 and 3500, respectively. Assume our typical person starts a diet with a daily caloric intake of d calories, and let x_i denote their weight after i days on the diet.

a. Set up the linear difference equation governing this diet.
b. Find the equation for x_n.
c. Solve this equation for d.
d. Use your formula to set the diet for a person weighing 185 lbs with constants $c = 15$ and $C = 3600$ who wishes to lose 11 lbs in 60 days.

[Hint for Exercise 3.13: It's true that when Jack weighs 176 lbs, a 1780-calorie diet would result in a daily deficit of 700 calories, contributing to weight loss. But what is the deficit when Jack's weight has dropped to 172 lbs? 165 lbs?]

The Mathematics of Personal Finance

4

4.1 Compound Interest

Suppose that you deposited \$100 in a bank account which pays 12% interest at the end of each year. At the end of the first year you will have in the bank your original \$100 plus $.12 \times \$100 = \$12$ in interest. If you keep the entire \$112 in the bank for a second year, then at the end of that second year you'll have \$112 plus $.12 \times \$112 = \$13.44$ in interest for a total of \$125.44. Suppose that you plan to leave the initial deposit and all interest in the bank for ten years. Can you compute the amount that you will have in the bank at the end of ten years without computing all of the intermediate values? One way of computing the amount at the end of one year is to simply multiply \$100 by (1.12):

$$\$100 + 0.12 \times \$100 = (1 + 0.12)\$100 = (1.12)\$100.$$

Then at the end of the second year, you will earn $.12 \times [(1.12)\$100]$ in interest so at the end of the second year you will have:

$$[(1.12)\$100] + 0.12 \times [(1.12)\$100] = (1.12)^2\$100;$$

and, similarly, after n years, you will have $(1.12)^n\$100$. This example helps us to see the general formula for compound interest.

Theorem 4.1 *Let P denote the principal, or the amount invested, let n denote the number of compounding periods and let i denote the rate at which interest is paid each period. Then F the future (or final) value of the investment is given by:*

$$F = (1 + i)^n P$$

© The Author(s), under exclusive license to Springer Nature Switzerland AG 2026
J. J. Edmond, J. E. Graver, *The Linear Function and Euclidean Geometry*, Synthesis
Lectures on Mathematics & Statistics, https://doi.org/10.1007/978-3-032-22712-6_4

Most of the time compounding is done more frequently than once a year; quarterly, monthly or even daily. Before high speed calculators and computers, compounding on savings accounts was usually done quarterly; now it is common to compound monthly or daily. Suppose we consider your initial investment: $100 at 12% for one year, but compound the interest quarterly. In our formula, $P = \$100$ and $n = 4$; but what is i? Since our periods are one-fourth of a year long the interest earned for this period would be $\frac{1}{4} \times 12\%$ or 3%. We have then:

$$F = (1.03)^4\$100 = (1.125508810)\$100 = \$112.55$$

So the effect of compounding quarterly, for one year, is to increase our yield by 55 cents on a $100 investment. Similarly, compounding monthly would yield $112.68 and compounding daily would give $112.75. You should verify these results on your own calculator. You might want to verify that more frequent compounding (each hour, each minute) won't increase your annual yield beyond. $112.75.

It is important to make a clear distinction between the annual rate of interest and the periodic rate. The standard convention is to report all interest rates as annual rates and then to compute the periodic rate from this. Thus, if you are told that you can invest savings at 5.75% or take a loan at 9.65%, it is understood that these rates are annual rates. In some instances where the annual rates are unseemly, they are reported differently. For example, your credit card company will probably state that their interest rate on the unpaid balance, computed monthly, is only $1\frac{1}{2}\%$. But, somewhere, in the fine print, they will admit that the annual rate is 18%.

Exercise 4.1 How much money will you have at the end of five years if you deposit $500 at 6.5% compounded monthly? How much more will you have if the compounding is daily?

Exercise 4.2 In answering each of the following questions, assume that the annual interest rate for your investment is 4.5% and that compounding is carried out monthly.

a. For how long must you leave $100 in the bank to accrue $1000?
b. For how long must you leave $1000 in the bank to accrue $10,000?
c. For how long must you leave $100 in the bank to accrue $10,000?
d. Can you give a "nice" explanation of what is going on?

Depositing $100 for one year at 12% compounded monthly yields $(1.01)^{12} \times \$100$ or $(1.126825) \times \$100$. That is, 12% compounded monthly has the same effect as an annual interest rate of 12.6825% without compounding or, as is usually stated, 12% compounded monthly has the same return as a simple interest rate of 12.6825%. This corresponding simple interest is often called the *effective rate* or *annual yield*. Certificates of deposit

(CD's) and other long term investments are often described by giving both their annual rate and their their annual yield.

Exercise 4.3 In each of the following cases, compute the annual yield:

a. an annual rate of 5.8% compounded quarterly
b. an annual rate of 5.8% compounded monthly
c. an annual rate of 5.8% compounded daily

There is a natural connection between frequent compounding and the natural constant e. This constant, like π, is irrational; the first few terms of its decimal expansion are $e = 2.71828182845\cdots$. The constant π is defined to be the ratio of the circumference of a circle to its diameter; the constant e is defined by a limit: Consider the function $h(x) = (1 + \frac{1}{x})^x$. One easily checks that $h(1) = 2$, $h(2) = 2.25$, $h(3) = 2.\overline{370}$ and $h(4) = 2.44140625$.

Exercise 4.4 Compute $h(10), h(100), h(1000), h(10{,}000), h(100{,}000)$ and $h(1{,}000{,}000)$.

You should observe that as we take x larger and larger, the values of $h(x)$ get closer and closer together. In fact they get closer and closer to the specific irrational number which we denote by e. We say that e is the limit of the function $h(x)$ as x goes to infinity and write: $\lim_{k\to\infty}(1 + \frac{1}{k})^k = e$.

Let a denote the annual rate for an investment and k the number of compounding periods per year. Then the formula in Theorem 4.1 may be rewritten as:

$$F = (1 + \frac{a}{k})^{ky} P,$$

where y is the length in years of the investment and $n = ky$ is the total number of periods of the investment. The reader who has studied limits should find the following computation familiar:

$$\lim_{k\to\infty}(1 + \frac{a}{k})^{ky} P = \left(\lim_{k\to\infty}(1 + \frac{a}{k})^{\frac{k}{a}}\right)^{ay} P = e^{ay} P$$

For the reader not familiar with limits, the next exercise should show that the exponential function e^{ya} is a good approximation to $(1 + \frac{a}{k})^{ky}$ when k is large. Thus, for very frequent compounding, the formula:

$$F = e^{ay} P$$

may be used to get a very accurate approximation. When using the exponential function to compute compound interest, we say that we are compounding *instantaneously* or *continuously*.

Exercise 4.5 Consider a $10,000 investment for 5 years at 4.8% interest and compute its value at the end of the five years if:

a. compounding is done quarterly
b. compounding is done monthly
c. compounding is done daily
d. compounding is done hourly
e. compounding is done continuously

Compound interest could be discussed in the context of linear difference equations: if i is the periodic rate, x_0 the amount of the initial deposit and x_n the amount in the account after n periods, then $x_n = (1 + i)x_{n-1}$ and, since b and c are both zero, we have, by the formula for the nth that $x_n = (1 + i)^n x_0$. Because of the ease in deriving the formula for compound interest directly, we really don't need to use the theory of linear difference equations. However as we move on to loans and annuities, we will make extensive use of the theory of linear difference equations.

4.2   Annuities and Loans

Having explored the concepts of simple and compound interest in basic loans and savings, we now turn to more complex financial products—such as annuities—which involve regular payments and play a key role in retirement planning and investment strategies.

Suppose that at the end of each month you deposit $100 in a bank account at an annual rate of 12% compounded monthly. How much will you have at the end of 5 years? Suppose that you deposit $10,000 in an account and withdraw $200 at the end of each month; how long will the money hold out? Suppose that you borrow $10,000 for 5 years at 12%; what is the size of your monthly payments? Any investment scheme or loan which involves regular payments is called an *annuity*. In developing a formula for annuities, there are two conventions that are commonly adopted: first, all payments or withdrawals are made at the end of each period; second, the compounding periods and the pay periods are the same.

Let us consider these senarios one at a time. First, we consider depositing $100 at the end of each month for 5 years without taking any out. Let x_n denote the amount in the account at the end of the nth or the start of the $(n + 1)^{\text{st}}$ month. Due to the convention that payments are made at the end of each period, $x_0 = 0$. At the end of the first month or start of the second, we have just deposited $100 but it has not yet accrued any interest; so $x_1 = 100$ (it is convenient to omit the $ in front of the values of the x_i). Now x_2 is x_1 plus the interest it has accrued $(1.01) \times x_1 = 101$ plus the new payment of 100 for a total of $x = 201$. In general, the amount at the end of the nth month, x_n, is equal to the amount at the start of the month, x_{n-1}, plus the interest it has generated $(.01)x_{n-1}$ plus the next payment of 100. Thus $\{x_n\}$ is the solution the linear difference equation:

$$x_n = (1.01)x_{n-1} + 100, \quad \text{with initial value } x_0 = 0.$$

Here $a = (1.01)$, $b = 100$ and $c = \frac{b}{1-a} = -1000$. So $x_n = (1.01)^n(1000) - 1000$ In a 5 year loan, $n = 60$, which yields \$8166.97. \$6000 of this amount represents the amount that you actually deposited, the remaining \$2166.97 is the interest earned.

Now consider scenario #2 where we start by depositing \$1000 in an account paying 12%, compounded monthly and we withdraw \$200 each month. This scenario also leads to a linear difference equation. As above, let x_n denote the amount in the account at the end of the nth or the start of the $(n+1)^{\text{st}}$ month. Here $x_0 = 10,000$ and x_1 is computed by adding the interest accumulated in the first month and then subtracting the withdrawal of \$200: $(1.01)10,000 - 200 = 9900$.

The equation for the nth balance is:

$$x_n = (1.01)x_{n-1} - 200;$$

$$\text{so } c = 20,000 \quad \text{and} \quad x_n = (1.01)^n(-10,000) + 20,000.$$

We must then solve the equation $(1.01)^n(-10,000) + 20,000 = 0$ for n to see when our funds will dry up. Simplifying, we have $(1.01)^n = 2$. This could be solved by "trial and error" or by using logs to get $n = 69.66$. Thus, you will be able to take \$200 a month from the account for 5 years and 9 months; in the month 70 you will withdraw somewhat less than \$200 and close the account.

Exercise 4.6 Considering the scenario we've been working through:

a. How much will you actually be able to withdraw in the month 70?
b. Over the 70 months, how much interest will you earn?

The last of the original three scenarios is similar to the second except that the amount of the payment is unknown. The total amount due initially is the amount you receive from the loan, in this case, \$10,000; so, $x_0 = 10,000$. At the end of one month, we compute the total amount due by adding in the interest and subtracting off the first payment: $x_1 = (1.01)x_0 - r$, where r denotes the payment size. (The payment on a loan is often called the "rent" in finance.) We see then that the loan is described by the following difference equation:

$$x_n = (1.01)x_{n-1} - r;$$

where r is the amount of each payments and $x_0 = 10{,}000$ is the amount of the loan. We compute: $c = \frac{-r}{1-(1.01)} = 100r$ and $x_n = (1.01)^n(10{,}000 - 100r) + 100r$.

Since the length of the loan is 60 months, $x_{60} = 0$. To determine what the monthly payment is, we wish then to solve $(1.01)^{60}(10{,}000 - 100r) + 100r = 0$ for r. Thus, $r = \frac{(1.01)^{60}(10{,}000)}{100((1.01)^{60}-1)} = 222.44$. Hence, each of the 60 payments on this loan are $222.44; giving a total of $13,346.40 in payments. $10,000 is the principal and $3346.40 is interest.

Now let's develop the general formula for annuities. Suppose that you deposit r dollars at the end of each period, for n periods at a periodic rate of i. Letting x_n denote the value of the investment at the end of the nth period, the linear difference equation which describes this is:

$$x_n = (1+i)x_{n-1} + r, \quad \text{where } x_0 = 0.$$

We compute c to be $\frac{r}{(1-(1+i))} = -\frac{r}{i}$; so x_n can be rewritten as $\frac{(1+i)^n-1}{i}r$.

Theorem 4.2   *The (future) value, F, of n regular periodic payments of $r at a periodic rate of i is given by:*

$$F = \frac{(1+i)^n - 1}{i}r.$$

Exercise 4.7   Upon entering the work force at age 20, Malcolm starts saving for his retirement at age 65. He opens an account at an annual rate of 5.6% compounded monthly into which he will make regular monthly payments.

a. If his payments are $100, how much will be in the account at the time of his retirement? How much of that amount represents interest earned?
b. How big should his payments be if he wishes to retire a millionaire?

Exercise 4.8   A manufacturer has just bought a piece of equipment for $500,000. She expects to have to replace this piece of equipment in ten years and wishes to set up a "sinking fund" to pay for the replacement, i.e., an annuity which, in ten years, will be worth the cost of the replacement.

a. What should be the size of her monthly payment into the sinking fund? In answering this question, use the following data: the manufacturer assumes that the cost of the replacement will go up by an annual inflation rate of 3% and that the investments in the annuity will grow at an annual rate of 4.5% compounded monthly. First compute the amount that the manufacturer assumes that he will need in ten years and then compute the payments that will achieve that target.
b. After five years the economic climate has changed and our manufacturer decides to adjust is payments into the sinking fund. She notes first that, at this time, the cost of

a replacement is \$600,000, not the \$579,637 predicted by a 3% inflation rate and that an inflation rate of 4% is expected for the next five years. The good news is that the funds now in the sinking fund along with all future payments will earn 6% compounded monthly. Find her new payments.

4.3   More About Loans

Next consider a loan charging interest at $100i\%$ per period with payments of r dollars. If we denote the amount due on the loan at the end of the kth period, the linear difference equation which describes a loan is:

$$x_k = (1+i)x_{k-1} - r,$$

where x_0 is the amount of the loan. We easily compute c to be $\frac{-r}{(1-(1+i))} = \frac{r}{i}$; so the value of x_k is:

$$x_k = (1+i)^k\left(x_0 - \frac{r}{i}\right) + \frac{r}{i}.$$

Let n denote the length of the loan, i.e., the number of periods or payments needed to pay off the loan. Then $x_n = 0$ since this is when the loan will be paid off. Setting $x_0 = L$ and collecting the terms involving $\frac{r}{i}$, gives the formula in Part (3) below. Then setting $k = n$ and $x_n = 0$ and solving for L, gives the formula in Part (1). Finally solving this equation for r, gives the formula in Part (2).

Theorem 4.3

1. If L denotes the amount of a loan in dollars to be paid off in n periodic payments of r dollars at a periodic rate interest i, then:

$$L = \frac{(1+i)^n - 1}{i(1+i)^n} r$$

2. If you borrow L dollars for n periods at a periodic rate i, your payments will be given by:

$$r = \frac{i(1+i)^n}{(1+i)^n - 1} L$$

3. If you borrow \$L with periodic payments of r at a periodic rate i, then, x_k, the amount due on the loan after k periods is given by:

$$x_k = (1+i)^k L - \frac{(1+i)^k - 1}{i} r$$

Exercise 4.9 Write out the proof of Theorem 4.3.

Exercise 4.10

a. Substitute the value for L given in part **a** of the theorem into the formula for x_k in part **c**. Then simplify to get:

$$x_k = \frac{(1+i)^{n-k} - 1}{i(1+i)^{n-k}} r$$

b. Interpret x_k as the amount that one could borrow for $n - k$ periods at a periodic rate of i with payment size r which will prove the following corollary.

Corollary 4.1 *The balance still due on a loan after k periods is equal to the amount you could borrow for $n - k$ periods at a periodic rate of i and payment size r where i, r and n are the periodic rate, rent and length of the original loan.*

Exercise 4.11 (Research Project) Before high speed calculators, the balance due on a loan was often computed by a rather simple computation called the "Rule of 78s". Look up this method, explain it and then discuss just how accurate it was.

Exercise 4.12 Carter wishes to buy a car, he has \$6500 for a down payment and the "going" interest rate is 10.5%.

a. How big will his payments be if he wishes to buy a \$23,500 car and pay it off in 4 years? How much in interest will he pay over the life of the loan?
b. If he can make monthly payments of only \$500, how expensive a car can he buy with a 4 year loan, a 5 year loan? In each case, how much interest will he pay over the life of the loan?

The difference equation approach to loans allows us to get far more information about the loan than just the size of the payments. Recall, that the linear difference equation for the balance due on a loan after k payments is:

$$x_k = (1+i)x_{k-1} - r$$

Recall that the fixed point for this difference equation is $\frac{r}{i}$. Since $a = (1+i) > 1$, the terms x_k diverge monotonically from $\frac{r}{i}$. If the initial loan value x_0 is less than $\frac{r}{i}$, successive values will decrease, and eventually, the loan will be paid off. Conversely, if x_0 is greater than $\frac{r}{i}$, successive values will increase, and the loan will never be paid off. The condition for eventual loan repayment is $x_0 < \frac{r}{i}$, which translates to $iL < r$. Here, iL represents the interest accrued on L dollars in one period. Therefore, if $r > iL$, the initial payment of r dollars will cover the interest and reduce the debt by $r - iL$ dollars. As the new debt is now less than L, subsequent payments will eventually clear the loan. However, if $r < iL$, the initial payment will not cover the interest, causing the debt to increase. In this case, the debt will grow with each payment, and the loan will never be paid off.

Consider a \$10,000 five year loan at 12%. Using the formula in Theorem 4.3b, we have that the monthly payments will be \$222.44. which is more than double the interest iL or \$100. Now compare this with a 30 year \$100,000 mortgage also at 12%. In this second case, iL is \$1000 while the payments compute to only \$1028.62! This is the major difference between short-term and long-term loans: A large part of all payments of a short-term go toward reducing the debt while the early payments on a long-term loan are almost all interest. Of the first payment of the mortgage just described, \$1000 went toward interest while the debt was reduced by only \$28.62.

To better understand this difference between short term and long term loans, let us consider two specific examples: a \$100,000, 25-year mortgage at 6% and a \$100,000, 5 year business loan also at 6%. The monthly payments for the mortgage are:

$$r = \frac{.005(1.005)^{300}}{1 - (1.005)^{300}} = 644.30.$$

while the monthly payments on the 5 year business loan are:

$$r = \frac{.005(1.005)^{60}}{1 - (1.005)^{60}} = 1933.28.$$

Since the interest on \$100,000 for one month at 6% is \$500, we see that only \$144.30 of the first mortgage payment goes to reducing the debt while the first business loan payment reduces the debt by \$1433.28. Just why this dramatic difference occurs can best be understood using our one-line graphs.

Consider an arbitrary loan in the amount L for n periods with a periodic rate of i and payments of r dollars per period. In the following figure below we have indicated the center ($\frac{r}{i}$) of the difference equation governing this loan. For this illustration we have taken $n = 11$.

$$
\begin{array}{c}
0 \quad\; x_9 \quad\; x_7 \;\; x_5 \; x_3 \, x_1 \\
\longmapsto\!\!+\!\!+\!\!+\!\!+\!\!+\!\!+\!\!+\!\!+\!\!+\!\!+\!\!+\!\!+\!\!+\!\!\longrightarrow \\
x_{11} x_{10} \quad\; x_8 \quad\; x_6 \; x_4 \, x_2 \, L \qquad\qquad\qquad \frac{r}{i}
\end{array}
$$

For a loan, the magnification of the associated similarity is $(1 + i)$. Thus the distance between the center and x_1 is $(1 + i)$ times the distance between the center and x_0 or the distance between x_1 and x_0 is I times the distance between the center and x_0. This helps to explains the fundamental difference between a long term and a short term loans. In a long term loan such as the mortgage described above, the value of the loan is relatively close to the center and the balance due does not decrease very much at the outset. For this mortgage, we have $\frac{r}{i} = \frac{644.30}{.005} = 128{,}860$ and some values for the balance due on this mortgage are sketched below. Of the first payment, $500 goes for interest while the balance due is reduced by only \$144.30. The value of x_1 (\$99,855.70) is so close to x_0 that it cannot be distinguished from x_0 on our graph. We have graphed the balance due in 5-year intervals: $x_{60} = \$89{,}932.20$, $x_{120} = \$76{,}352.21$, $x_{180} = \$58{,}034.86$ and $x_{240} = \$33{,}327.50$.

For the 5 year business loan we have $\frac{r}{i} = \frac{1933.23}{.005} = \$386{,}656$. Of the first payment, \$500 goes for interest while the debt is reduced by \$1433.23. So debt reduction is more significant from the very start and the balance due values do not bunch up as much about x_0. We have graphed the balance due each year: $x_{12} = \$82{,}319.69$, $x_{24} = \$63{,}548.84$, $x_{36} = \$43{,}620.35$ and $x_{48} = \$22{,}462.66$.

This method of picturing a loan has one other useful feature. If we make a simple change of scale, we have a geometric representation of the distribution of each payment between interest and debt reduction:

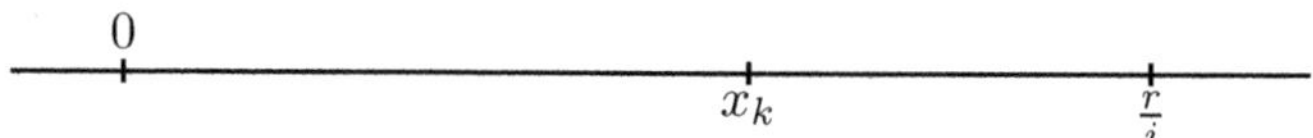

Multiplying through by i, we have:

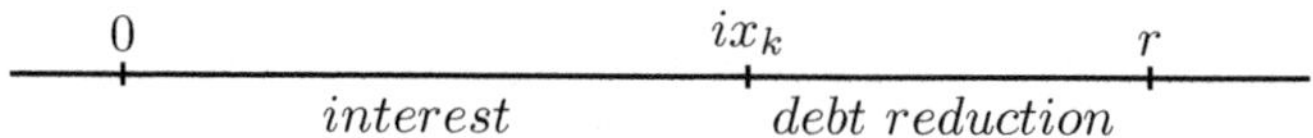

Of the $(k+1)^{\text{st}}$ payment of r dollars, ix_k dollars is the interest on the previous balance while the debt is reduced by $r - ix_k$ dollars.

Exercise 4.13 Upon entering the work force, Tamsin starts saving for her \$55,000 dream car. She opens an account at an annual rate of 4.6% compounded monthly into which she will make regular monthly payments. She also has the option of borrowing at a 10.2% annual rate.

a. How big should her payments be if she wishes to have the $55,000 for the car in 3 years?
b. If she borrows the full $55,000 to buy the car now and pays off the loan in 3 years, how big will her payments be?
c. Tamsin decides to save for 18 months, then buy the car and pay off the rest with a loan over the next 18 months. She wants her payments into the savings account and her loan payments to be about the same. What will the payments be?

Exercise 4.14   Ten years ago Kaia bought a house taking out a 25 year $275,000 mortgage at 8.6%.

a. How big are her payments.
b. Now, after exactly 10 years, how much does she still owe?
c. For a flat fee of $3000 (which can be added to the loan) she can refinance the loan at 6.8%, resulting a 15 year mortgage for the balance due plus $3000. If she was to do this what would her new payments be?

Loans can come in various configurations. For instance, with a credit card or a home equity loan, each payment includes both the total interest accrued and a fixed percentage of either the original loan amount or the remaining balance. As a result, early payments will be larger compared to those of a traditional loan and will decrease over time. Additionally, the total interest paid will generally be less than that of a traditional loan, assuming the same interest rate for both types of loans. To illustrate this, let us examine the specific case of Kaia's house purchase.

Exercise 4.15   Kaia paid off her home mortgage 5 years ago. Now she wants to buy a new car and to pay for it she has taken out a $15,000 home equity loan at 9.5% interest to be paid off in monthly payments over 5 years. Each monthly payment consists of $\frac{1}{60}$ of the original loan amount, plus interest on the remaining balance.

a. How much is her first payment?, her second payment?, her third payment?, her last payment?
b. What is the total interest that she will pay?
c. If she can get a standard 5 year car loan at the same rate, what would her payments be?
d. What is the total interest paid with the standard loan?
e. Explain the difference in the amounts of interest paid by the two schemes.

Linear Functions in Business

5

5.1 Supply and Demand

Supply and demand are fundamental concepts in economics that explain how prices and quantities of goods are determined in a market. Demand refers to the quantity of a product that consumers are willing and able to buy at various prices. Generally, as the price of a product decreases, the quantity demanded increases, and as the price increases, the quantity demanded decreases. Supply represents the quantity of a product that producers are willing and able to sell at different prices. Typically, as the price of a product rises, the quantity supplied increases, and vice versa.

The interaction between supply and demand determines the market equilibrium price and quantity. At equilibrium, the amount of the product that consumers want to buy equals the amount that producers want to sell, leading to a stable market situation. Changes in factors like consumer preferences, production costs, or external events can shift supply and demand curves, affecting prices and quantities in the market.

The *demand function* for a product typically represents the sale price p as the dependent variable and the quantity produced q as the independent variable. Consider the following demand function for a small appliance:

$$p = -0.0004q + 160.$$

If 100,000 appliances are produced the price will be $120.00; if the quantity manufactured is doubled to 200,000, the price will drop to $80.00. Like most mathematical models for "real life" problems, this model only makes sense within a certain range. For example, if a half a million appliances are produced the price will *not* fall to $40.00. Let us assume that our model is reasonably accurate in the production range from 50,000 to 250,000 appliances.

© The Author(s), under exclusive license to Springer Nature Switzerland AG 2026
J. J. Edmond, J. E. Graver, *The Linear Function and Euclidean Geometry*, Synthesis
Lectures on Mathematics & Statistics, https://doi.org/10.1007/978-3-032-22712-6_5

The *supply function* for the product is also represented with p as the dependent variable and q as the independent variable. Consider the following supply function for the same small appliance:

$$p = 0.0005q + 50,$$

which we assume to be valid on the same production range. Thus, 100,000 appliances will be produced for an expected price of $100.00 while the production level will be set at 200,000 if the expected sales price is $150.00. If, on the other hand, the price drops to $75 per appliance, only 50,000 will be produced.

Exercise 5.1 Typically the supply equation is well-known to the manufacturer. However, the demand equation is constructed from experience and market research. Initially, the manufacturer believes that the appliance will sell well at a price of $100.

a. How many appliances will the manufacturer produce in this first month? How many of these will he be able to sell?
b. Based on his experience this first month he decides to produce 120,000 appliances in the second month. At what price will he sell them? How many of these will he be able to sell?
c. Based on his experience in the second month he decides to produce 130,000 appliances in the third month. At what price will he sell them? How many of these will he be able to sell?
d. After several months of experimentation, he guesses that consumer demand is described by the linear (demand) equation given above. If he wishes to make just enough appliances so that they will all sell while meeting the full demand, how many will he produce per month? At what price will he sell them?

Exercise 5.2 Suppose that the supply equation of another appliance is given by $p = .006q + 103.5$ and assume that the manufacturer collects the following data: In the first month, he produces 5000 appliances and sells them all at $133.50 and believes that he could have sold over 4000 more. In the second month, he produces 7000 and again sells them all, this time at $145.50; also this time he has an additional 786 unfilled orders. In the third month, he produces 8000 and sells them at $151.50. But, this month he has 1071 left over. Is there a linear demand equation consistent with the above data? If so, find it and compute the number of appliances the manufacturer should produce a month. If there is no linear demand equation, explain why there is none.

As the above exercises indicate, the theory behind this supply and demand model is that, in the open marketplace, the price and production level will stabilize at the point of intersection of the supply and demand curves. For our original model, this occurs at the solution to the system:

$$p = -0.0004q + 160$$

$$p = 0.0005q + 50$$

One easily checks that the solution to the system gives a production level of 122,222 appliances and a price of \$111.11 per appliance. A second justification of this theoretical expectation is based on the assumption that the price will be set by the market place and not the manufacturer. Again suppose that the time frame for this model is one month and that in the first month 100,000 appliances are produced and sold. Because of the high demand, the selling price rises to \$120 ($p = -0.0004(100{,}000) + 160 = 120$). Responding to this price, the manufacturer produces 140,000 appliances in the second month ($q = \frac{1}{0.0005}(120 - 50)$). But now the price drops to \$104.00 ($p = -0.0004(140{,}000) + 160$). In response to the drop in price, the third month's production level is set at 108,000 ($q = \frac{1}{0.0005}(104 - 50)$) resulting in a price increase to \$116.80, etc. This iterative process is illustrated in the following diagram:

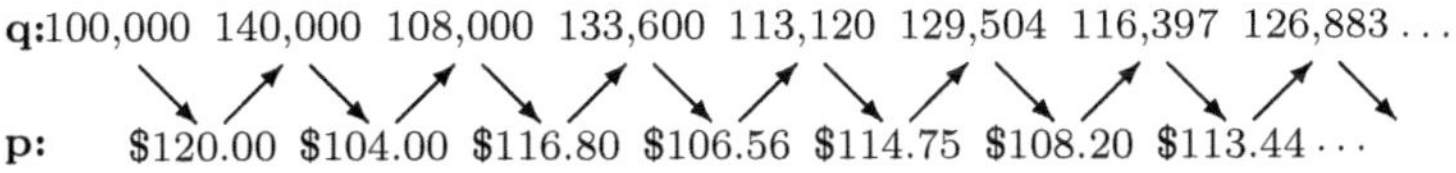

This process will eventually stabilize at $q = 122{,}222$ and $p = 111.11$. It is not surprising that this iterative process may be described by a linear difference equation. Let q_n denote the level of production in the $(n-1)^{\text{st}}$ month. Then the price in the $(n-1)$st month would be $-0.0004q_n + 160$ and, as a result of this new price, the production level set for the next month, q_{n+1}, would be set at $\frac{1}{0.0005}((-0.0004q_{n-1} + 160) - 50)$. Simplifying, we get:

$$q_n = -\frac{4}{5}q_{n-1} + 220{,}000$$

which has fixed point:

$$c = \frac{220{,}000}{1 + \frac{4}{5}} = 122{,}222.222 \cdots$$

Making up interesting examples that have reasonable solutions is always a problem for the teacher and textbook writer. One useful technique is to solve the problem symbolically and then choose the constants to give a nice example. For example, linear demand and supply equations always have the form:

$$p = -dq + e \quad \text{and} \quad p = sq + t, \quad \text{where } d, e, s, t \text{ are all positive constants.}$$

Exercise 5.3

a. Solve the above system for p and q in terms of the constants d, e, s, and t.
b. Create a supply and demand model for a business firm that has equilibrium values of $q = 35{,}000$ and $p = 7.25$. Also, construct your models so that the supply and demand equations make sense in the production range $5000 \leq q \leq 60{,}000$. Finally, try to construct your model so that the prices for both equations do not fluctuate more than \$2.00 in the given range.

5.2   Supply and Demand with Taxes

Supply and demand equations can be used to evaluate how taxes impact the production level and sales price of a product. There are two primary types of taxes: a *flat tax* of a fixed amount or a *percentage tax*, a fixed percentage of the sale price. Also the tax may be levied on the consumer as a *sales tax* or on the manufacturer as a *manufacturer's tax*. This results in four possible tax configurations. Regardless of the type, the key considerations for the government when deciding to implement a tax remain the same:

- What will be the total revenue generated by the tax for the government entity imposing it?
- How much will the price of the product change?
- How much will the production level change?

The significance of the last question arises from its secondary effects. A substantial reduction in production could lead to job losses and decreased wages for workers, which, in turn, would result in lower tax revenue. Unemployed workers would contribute less in taxes and may require public assistance.

Let's examine how taxes would apply to the production and sale of the small appliance we discussed earlier. To compare all four options, consider a 5% tax on the price of small appliance or a flat tax of \$5.56 (0.05×111.11). The important thing to keep in mind as we consider the various options is that the variable p must always represent the actual amount paid by the consumer to the manufacturer and q represents the number of items manufactured and sold.

We start with the flat sales tax. Consider our example with a flat \$5.56 sales tax, i.e. \$5.56 to be paid by the consumer on top of the sale price p. Since the manufacturer is not involved in the taxation (except that he may collect it for the government), the supply equation remains the same. On the other hand, the number of appliances that will be bought will depend on the total amount paid by the consumer not simply p, the amount paid to the manufacturer. This leads us to need to alter the demand equation. The price p is replaced, in the demand equation, by $p + 5.56$, the amount that the consumer actually pays. Solving for p, then yields a new demand equation:

$$(p + 5.56) = -0.0004q + 160 \quad \text{or} \quad p = -0.0004q + 154.44.$$

Solving simultaneously the supply equation and this new demand equation gives a production level of 116,044 appliances at a price of \$108.02 each. The actual price paid by the consumer will be \$113.58 (108.02+5.56). The government will collect \$645,204.64 in taxes per month (5.56 × 116,044). However, the level of production will drop by 5% or 6178 appliances per month to be exact.

Now suppose that the flat \$5.56 tax is imposed as a manufacturer's tax, i.e. \$5.56 to be paid by the manufacturer for each appliance sold. Since this time the consumer is not involved in the taxation, the demand equation remains the same and it is the supply equation which must be changed. The number of appliances that will be produced depends on the portion of amount paid by the consumer which will be retained by the manufacturer. The new supply equation is obtained by substituting $p - 5.56$ for p in the supply equation and solving for p:

$$(p - 5.56) = 0.0005q + 50 \quad \text{or} \quad p = 0.0005q + 55.56.$$

Solving the new supply equation and the original demand equation simultaneously gives a production level of 116,044 appliances at a price of \$113.58 each. Again, the government collects \$645,204.64 in taxes per month and the level of production drops by 6178 appliances per month. It seems that whether a flat tax is imposed as a sales tax or a manufacturer's tax is immaterial, except, possibly, for political considerations. We will see shortly that this is not the case for a percentage tax.

Exercise 5.4 On the same set of axes, plot the original and the new supply and demand equations. Explain geometrically why a flat tax has the same effect on production levels whether it is imposed on the consumer or the manufacturer.

Now we consider a 5% sales tax, i.e. a tax of $0.05p$ to be paid by the consumer in addition to the sale price p. Again the manufacturer is not involved in the taxation and the supply equation remains the same. As noted above, the number of appliances that will be sold will depend on the total amount paid by the consumer and we must alter the demand equation by substituting $p + (.05)p$ for p:

$$(p + .05p) = (1.05)p = -0.0004q + 160 \quad \text{or} \quad p = \frac{-0.0004}{1.05}q + \frac{160}{1.05}.$$

Solving simultaneously the supply equation and this new demand equation gives a production level of 116,216 appliances at a price of \$108.11 each. The actual price paid by the consumer will be \$113.52 (1.05 × 108.11). The government can expect to collect \$628,728.56 in taxes per month. However, the level of production will drop by 6006 appliances per month.

Finally, we consider a 5% manufacturer's tax. As before, the demand equation remains the same and it is the supply equation which must be changed. The new supply equation is:

$$(p - (0.05)p) = (0.95)p = 0.0005q + 50 \ \text{ or } \ p = \frac{0.0005}{0.95}q + \frac{50}{0.95}.$$

Solving the new supply equation and the original demand equation simultaneously gives a production level of 115,909 appliances at a price of \$113.64 each. this time, the government collects \$658,581.94 in taxes per month and the level of production will drop by 6313 appliances per month. (In computing the total tax collected for the sales tax, 0.05×108.11 is rounded first and then multiplied by 116,216; no rounding takes place when the total manufacturer's tax is computed, since the manufacturer is not paying the tax appliance by appliance.)

In the case of a percentage tax, the price to the consumer is much the same whether the tax is a sales tax or a manufacturer's tax. From the government point of view, the manufacturer's tax produces significantly more in revenue. But, on the down side, the manufacturer's tax results in a significantly lower production level. How does the manufacturer, as opposed to their employees, fare under these tax schemes? To address this question, we need to examine revenue, production costs, and profits, which will be our next area of focus.

Exercise 5.5 On the same set of axes, plot the original and the new supply and demand equations. Explain geometrically why a percentage tax behaves differently when imposed on the consumer and the manufacturer.

Exercise 5.6 Rework these examples with a 9% tax or a flat \$10.00 tax.

Exercise 5.7 Consider the general model discussed in Exercise 5.3 **a**.

a. Compute the new equilibrium point when a flat tax of f is imposed on the consumer.
b. Compute the new equilibrium point when a percentage tax of f is imposed on the consumer.

5.3   Revenue, Production Costs and Profits

For our small appliance manufacturer, we can explore the company's potential profits under different conditions. Profit is calculated as revenue (the total amount earned) minus costs. Therefore, the initial step in analyzing profits is to assess revenue. Revenue itself is determined by multiplying the selling price by the quantity sold: $p \times q$, keeping the same definitions for those variables as in the previous section. In our original example, the manufacturer produced and sold 122,222 appliances per month at \$111.11 each.

The manufacturer's revenue is therefore \$13,580,086.42 per month. With a \$5.56 sales tax the revenue is about one million less, \$108.02 $\times$ 116,044 $=$ \$12,535,072.88, to be exact. If this flat tax is imposed as a manufacturer's tax, we compute \$113.58 $\times$ 116,044 $=$ \$13,180,277.52. But, subtracting the \$645,204.64 in taxed passed directly to the government, we again get \$12,535,072.88, which should come as no surprise. The one million dollar reduction in revenue due to the tax is counterbalanced by a decrease in production levels and, presumably, lower production costs. At this stage, it remains unclear how the tax will ultimately affect profits.

Exercise 5.8 Compute the revenues for our small appliance manufacturer when a 5% sales tax or manufacturer's tax is imposed.

It is very natural to expect a linear model for production costs. Production costs often fall into two categories: *fixed costs* and *variable* or *marginal costs*. Examples of fixed costs include rent on the factory, equipment maintenance costs and many management costs. Examples of variable costs include raw materials, equipment operating costs and most labor costs. Fixed costs are reported as a single figure while marginal costs are reported as a per item amount. Suppose that it costs \$755,000 a month in fixed costs to keep our small appliance factory running and that raw materials, labor costs and other variable costs come to \$101.50 per appliance. The resulting cost function is then: $c = c(q) = (101.5)q + 755,000$. Now that we have the cost function we may now compute the company's profits under the various tax schemes discussed above. Before a tax was imposed, the monthly revenue was just computed to be \$13,580,086.42. The monthly costs are $c(122,222) = 101.5 \times 122,222 + 755,000 = 13,160,533$. Thus, if no tax is imposed, the company will show a profit \$419,553.42 each month. If the flat tax (sales or manufacturer's) is imposed, revenue was computed above to be \$12,535,072.88 per month. The monthly production cost at the new lower production level is $c(116,044) = 12,533,466$. The result is a dramatic drop in profit to only \$1616.88 per month!

Exercise 5.9 For this small appliance manufacturer:

a. Compute the monthly production cost and profit for this company when a 5% tax is imposed. Which is better for the manufacturer a 5% sales tax or a 5% manufacturer's tax?
b. Without computing, what do you think the effect of a 9% or \$10.00 tax would be no matter how it is imposed? Check out your guess.

Exercise 5.10 Consider our small appliance factory and assume that a 5% sales tax is in effect. Suppose the manufacturer, Russell, has \$1,000,000 to invest in his company. He has three options: the \$1,000,000 could be used to relocate to a newer and smaller factory with lower rent and easier maintenance; it could be used to buy new robotic equipment on the production line; or it could be used to hire a lobbying firm to try to convince the government to reduce the present 5% manufacturer's tax. He believes that the move to the new location would result in reducing his monthly fixed costs by $250,000, that the new equipment on the production line would reduce his marginal costs by $2.10, and that the lobbying firm could get the tax reduced from 5% to 2.5%. Which option would result in the best financial gain for this factory?

5.4   Marginal Revenue and Marginal Profit

As we have described, the demand equation is an external factor for the manufacturer, meaning it is beyond his control. Similarly, without altering his operations, the cost function remains unaffected by the manufacturer. In contrast, the supply function is created by the manufacturer himself. In this section, we will explore the analysis involved in developing a supply equation. An essential aspect of any business operation is profit, and maximizing profit is a primary concern for the manufacturer.

Maximizing a function is usually considered a problem in the domain of calculus. However, in business mathematics, marginal functions often serve as an alternative to calculus. Suppose that we have a profit function $P(q)$ where, as above, q represents the quantity of items. Suppose that 1000 items are sold. Then the marginal profit at the production level 1000 is the *additional* profit that would be made if one more item were sold: $f(1001) - f(1000)$. Of course, we can compute the marginal profit at any production level: $f(q + 1) - f(q)$. This results in a new function of the variable q, $f^*(q) = f(q + 1) - f(q)$, called the marginal profit function. Of course, in this computation, the fact that f is a profit function has no significance and we may define the marginal function f^* for any function f.

In our manufacturing example, the cost function is linear, so the marginal cost is simply the slope of this function's graph. This is true of any linear function $f(q) = aq + b$:

$$f^*(q) = f(q + 1) - f(q) = (a(q + 1) + b) - (aq + b) = aq + a + b - aq - b = a.$$

The marginal cost should be interpreted as the additional cost of producing *one more* appliance. Note that the marginal cost is *not* the cost of producing *one* appliance. The cost of producing one appliance is called the *average* cost; it is denoted by $\bar{c}$; it is given by $\bar{c} = \bar{c}(q) = \frac{c(q)}{q}$ and it depends on the number of appliances produced. The average cost of producing one item is an important bit of information for the overall management of the company. However, for the specific decision of whether to increase, maintain, or decrease the current production level, the marginal cost is the most crucial factor.

Exercise 5.11   Let c be the linear cost function $c(q) = aq + b$.

a. Compute the average cost function $\bar{c}(q) = \frac{c(q)}{q}$. How are the marginal and average costs related for a linear cost function?

b. In the case of our small appliance factory, graph the marginal and average cost functions on the same axes.

When $f(q)$ is a linear function of q, then $f^*(q)$ is a constant equal to the slope of the graph of $f(q)$. When $f(q)$ is not a linear function, then $f^*(q)$ is not a constant function. Nevertheless, $f^*(q)$ can be interpreted as a slope. It is the slope of the line through the points $(q, f(q))$ and $(q + 1, f(q + 1))$: $\left(f^*(q) = \frac{f(q+1)-f(q)}{(q+1)-q} \right)$. In many cases, $f^*(q)$ is very close to the slope of the tangent line to the graph of $f(q)$ at the point $(q, f(q))$. In fact, if the graph of the function $f(q)$ is "smooth", then $f^*(q)$ is the slope of the tangent line to the graph of $f(q)$ at some point between $(q, f(q))$ and $(q + 1, f(q + 1))$. This follows from an important theorem from calculus called "The Mean Value Theorem". If you draw a few examples for yourself you will see that this is not a surprising result.

Exercise 5.12 Let $f(q) = \frac{2}{5}q^2 - q - 1$

a. Compute $f^*(q)$.
b. Graph $f(q)$, graph the tangent to this curve at $(q, f(q))$ and graph the line through $(q, f(q))$ and $(q + 1, f(q + 1))$ when $q = 2$.
c. Repeat **b** for $q = 3$ and $q = 4$.

Let us compute the marginal revenue and marginal profit for our small appliance manufacturer. As we have already noted, revenue is simply price times quantity sold. If q items are put on the market in a given month, the price at which they will all sell is determined by the demand equation. Hence, the revenue function is:

$$r = r(q) = pq = (-0.0004q + 160)q = -0.0004q^2 + 160q.$$

In the Table 5.1 we compute the revenue at several production levels:
Revenue seems to peak at about the production level of 200,000. We can verify this by considering the marginal revenue:

$$r^*(q) = r(q + 1) - r(q)$$
$$= \left(-0.0004(q + 1)^2 + 160(q + 1) \right) - \left((-0.0004q^2 + 160q) \right)$$
$$= -0.0008q + 159.9996.$$

Table 5.1 Revenue as a function of production quantity

Quantity	50,000	100,000	150,000	200,000	250,000
Revenue	$7,000,000	$12,000,000	$15,000,000	$16,000,000	$15,000,000

Table 5.2 Revenue and profit as functions of production quantity

Quantity	50,000	100,000	150,000	200,000	250,000
Revenue	$7,000,000	$12,000,000	$15,000,000	$16,000,000	$15,000,000
Profit	$1,170,000	$1,095,000	-$980,000	-$5,055,000	-$11,130,000

Rounding this to the nearest cent, gives $r^*(q) = -0.0008q + 160$. At $q = 100,000$, we have a marginal revenue of $80. This means that by producing one more appliance, 100,001 in all, the revenue will increase by $80. At $q = 150,000$ the marginal revenue has dropped to $40, at $q = 200,000$ it is zero and at $q = 250,000$ it is -$40. In this last case, the production of one more appliance will result in a lost of revenue. In general, if $q < 200,000$ and production is increased, revenue will increase while if $q > 200,000$ and production is increased, revenue will decrease. Thus, revenue can be maximized by setting the production level at 200,000.

Now let's turn our attention to the profit function. The profit function is defined to be revenue minus cost; so, the profit function is:

$$P = P(q) = r(q) - c(q) = (-0.0004q^2 + 160q) - (101.50q + 755,000)$$

$$= -0.0004q^2 + 58.50q - 755,000.$$

(We use capital P for profit to distinguish if from the price p.)

We will now add the profit information to Table 5.1 as shown in Table 5.2:
When revenue is at its highest ($16,000,000 at $q = 200,000$), the company will have a loss of over $5,000,000. The reason for this is clear: at this production level the sale price of each appliance is only $80 while it costs slightly more than $101.50 to produce that appliance! Profit seems to peak when production is low. To find out exactly where profit is maximum we compute the marginal profit. The easy way to do this is to observe that since profit equals revenue minus cost, marginal profit equals marginal revenue minus marginal cost. We'll use this formula now and you will justify it later:

$$P^*(q) = r^*(q) - c^*(q) = (-0.0008q + 160) - (101.50) = -0.0008q + 58.50.$$

One easily checks that the cutoff production level is 73,125: if fewer than 73,125 appliances are produced, marginal profit is positive and producing another appliance will increase profit; while, if more than 73,125 appliances are produced, marginal profit is negative and producing one more appliance will decrease profit. In this last case, producing one less appliance will usually increase profit. We conclude that the maximum profit possible is $1,383,906.25 and that will occur when 73,125 appliances are produced.

At this point it is quite natural to ask why this analysis yields a different production level and price from the one obtained using the supply and demand model. Since we are using the same demand equation, the explanation of this discrepancy must involve the supply equation. The fact is that the supply equation is the result of several different considerations, immediate profits being only one of them. There may be several reasons why a company may wish to increase production levels beyond that which maximizes profits. For example, maintaining its market: At a production level of 73,125 appliances the price is \$130.75. At this high a price another manufacturer may be tempted to start producing this appliance. Suppose that a competitor does enter the market and produces 49,097 appliances a month. Since 122,222 appliances are produced the price drops to \$111.11. Our manufacturers revenue drops to \$8,124,918.75; while his production costs drop to \$8,177,187.50. The result would be a monthly loss of \$52,268.75. Other considerations that may go into the supply equation include contractual agreements with employees and the size of the physical plant.

Exercise 5.13 Consider the demand equation discussed in Exercise 5.3a. and consider the general linear cost function $c(q) = mq + f$, where m is the marginal cost and f the fixed cost.

a. Compute the revenue function.
b. Compute the profit function.
c. Compute the marginal profit function.
d. At what production levels will the profit be maximized?
e. Select values for m and f which are consistent with the example you produced in Exercise 5.3 b. Arrange it so that your company makes a profit and so that the maximum profit occurs in the production range of the model.

We conclude this section with exercises on calculating and interpreting marginal functions. Readers with a background in calculus may notice the similarity between this method of calculating marginal functions and the calculus of derivatives.

Exercise 5.14 Let $g(q)$ and $h(q)$ be two functions of the quantity q.

a. Show that, if $f(q) = g(q) - h(q)$, then $f^*(q) = g^*(q) - h^*(q)$.
b. State and prove the analogous formula for f^* when $f = g + h$.
c. Show that, if $f = g \times h$, then

$$f^*(q) = h(q + 1) \times g^*(q) + h^*(q) \times g(q) \text{ and}$$

$$f^*(q) = h(q) \times g^*(q) + h^*(q) \times g(q + 1).$$

d. Show that, if $f = \frac{g}{h}$, then

$$f^*(q) = \frac{g^*(q) \times h(q) + g(q) \times h^*(q)}{h(q + 1) \times h(q)}.$$

Part II

The Two-Dimensional Linear Function

6.1   The Real Numbers

As we have seen, the real numbers are the heart of this book. We visualize them as lying along a number line. The points making up this infinite line are the real numbers evenly spaced, often labeled at the integers. The labels of other numbers can then be added as needed, e.g. $-\frac{3}{2}$ and $\sqrt{3}$. Without labels, the number line is a model of one-dimensional Euclidean geometry: its points are the points of the Euclidean line. This line on its own is homogeneous, meaning all points are treated the same.

In the next sections, we explore the Euclidean plane. Like the Euclidean line, the plane is homogeneous until a coordinate system is introduced. On the line, a coordinate system is obtained by choosing two points: one to represent the number 0, and another to represent 1. We usually place 1 to the right of 0 and take the segment [0,1] to define the unit length.

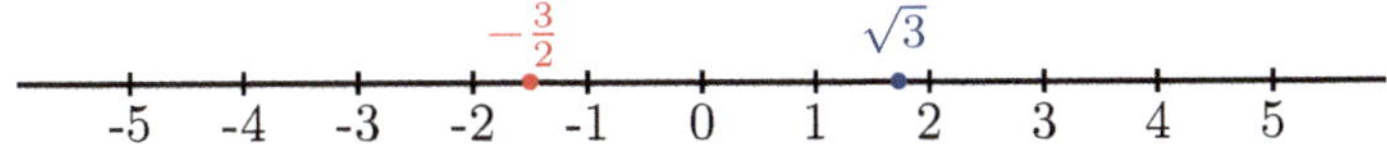

Now we can build the number line geometrically, as we picture it above with the integer points labeled. Starting at 1, we label the point one unit to its right as 2, then the point one unit to the right of 2 as 3, and so on. In the other direction, starting at 0 and moving left by unit lengths, we label the points as $-1, -2, -3$, and so on, creating the negative integers. To fill in the rest of the real numbers—fractions, decimals, and irrationals—we embed the number line as the x-axis in the Euclidean plane with the standard coordinate system.

To construct the point on the x-axis with coordinate $-\frac{3}{2}$, first construct the segment joining -1 on the x-axis and 2 on the y-axis. Then construct a line parallel to this segment that intersects the y-axis at 3. This line will intersect the x-axis at $-\frac{3}{2}$. To construct the

© The Author(s), under exclusive license to Springer Nature Switzerland AG 2026

J. J. Edmond, J. E. Graver, *The Linear Function and Euclidean Geometry*, Synthesis Lectures on Mathematics & Statistics, https://doi.org/10.1007/978-3-032-22712-6_6

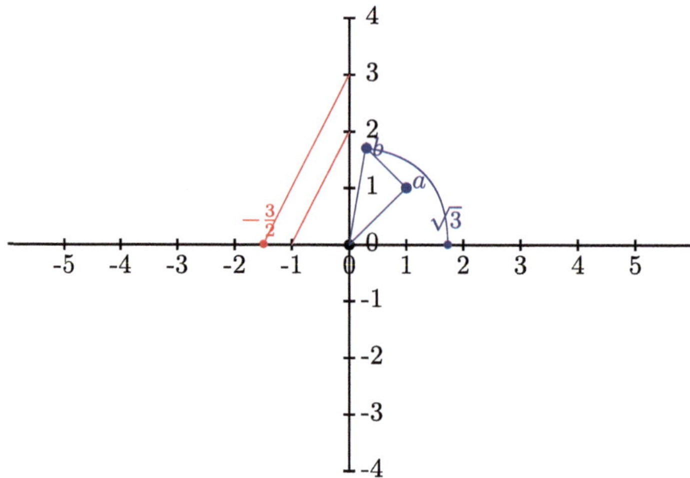

Fig. 6.1 Constructions for locating numbers

point on the x-axis with coordinate $\sqrt{3}$, construct the segment from the origin O to the point a with coordinates $(1, 1)$, and note that the segment $\overline{Oa}$ has length $\sqrt{2}$. Next, construct the line perpendicular to the segment $\overline{Oa}$ at the point a and mark off the segment $\overline{ab}$ of length 1. By the Pythatogean Theorem the segment $\overline{Ob}$ has length $\sqrt{3}$ and the circle with radius $\sqrt{3}$ centered at 0 passes through $\sqrt{3}$ on the x-axis.

The story of how the real numbers are constructed—starting with the natural numbers and building upward: first the rational numbers and then the irrational numbers—is a fascinating story, involving beautiful mathematics! We could provide a list of references here, but instead, we present it as a research project for the interested reader.

Project 6.1 Do your own research to find suitable papers or books on the construction of the real numbers.

Exercise 6.1 Using the red construction in Fig. 6.1, construct the points on the real line for the real numbers $\frac{4}{5}$ and $\frac{5}{4}$. Describe the process for locating any rational number.

Exercise 6.2 Using the blue construction in Fig. 6.1, construct the points on the real line for the real numbers $\sqrt{5}$ and $-\sqrt{6}$.

6.2   Basics of the Euclidean Plane

In the previous chapters, we investigated the linear functions that mapped $\mathbb{R}$ to $\mathbb{R}$ by analyzing their effects on the one-dimensional number line. We will now broaden our focus to "linear functions" that map $\mathbb{R}^2$ to $\mathbb{R}^2$, which we will call *two-dimensional linear functions* or *2D linear functions*, and explore their effects on the two-dimensional plane. To define and fully understand these functions, we must first understand the basic structure of the Euclidean plane. While in one-dimensional space we dealt only with points and the distance between them, two-dimensional space introduces lines in addition to points, allowing us to measure both distances between points and angles between lines.

In this context, we consider points in the plane as *vectors* and denote them using lowercase letters with a vector sign over them, such as $\vec{v}$. A vector is a mathematical object that has both direction and magnitude. It can be thought of as an arrow pointing from one location to another within the plane. If we assume the starting point of the vector is the origin then any vector can be described using coordinates within a standard coordinate system, which allows us to precisely specify the position of its endpoint relative to the origin and its direction in the plane.

To build this framework for $\mathbb{R}^2$, we establish the *standard coordinate system* consisting of:

(1) A point known as the *origin* or *zero vector*, denoted by $\vec{0}$.
(2) A horizontal line through the origin, called the *x-axis*, and a vertical line perpendicular to the *x*-axis at the origin, known as the *y-axis*.
(3) A linear scale on the *x*-axis where the origin is assigned 0 and the point one unit to the right is assigned 1. The vector from the origin to the point 1 on the *x*-axis is called the unit vector on the *x*-axis and is denoted by $\vec{x}$.
(4) A linear scale on the *y*-axis where the origin is assigned 0 and the point one unit above is assigned 1. The vector from the origin to the point 1 on the *y*-axis is called the unit vector on the *y*-axis and is denoted by $\vec{y}$.

Consider a point $\vec{v}$ in the Euclidean plane. To determine the coordinates of $\vec{v}$, first, drop the perpendicular line from the point $\vec{v}$ to the *x*-axis. The coordinate on the *x*-axis of the point where this line intersects the *x*-axis determines is called the *x*-coordinate of $\vec{v}$. Next, draw the line from $\vec{v}$ perpendicular to the *y*-axis. The coordinate on the *y*-axis of the point where this line intersects the *y*-axis determines is callled the *y*-coordinate of $\vec{v}$. Using the coordinates of $\vec{v}$, we can express $\vec{v}$ as a *row vector* (v_x, v_y) or as a *column vector* $\begin{bmatrix} v_x \\ v_y \end{bmatrix}$.

Additionally, $\vec{v}$ can be written as the vector sum $\begin{bmatrix} v_x \\ v_y \end{bmatrix} = v_x \begin{bmatrix} 1 \\ 0 \end{bmatrix} + v_y \begin{bmatrix} 0 \\ 1 \end{bmatrix}$ or simply $\vec{v} = v_x \vec{x} + v_y \vec{y}$:

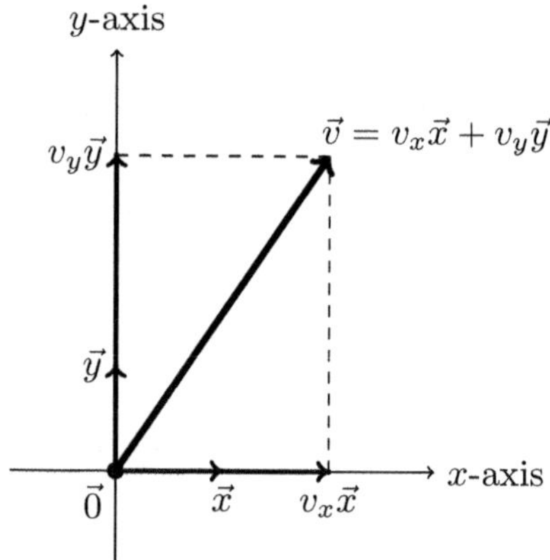

6.3   Transforming the Coordinate Grid

The coordinate system for vectors works similarly to the familiar standard coordinate system for points in the plane, providing a consistent framework for describing locations and directions. In this system, we have a *coordinate grid*, which consists of a network of vertical and horizontal lines that intersect the axes at points with integer coordinates. In the diagram showing $\vec{v}$ below, we have superimposed the standard coordinate grid. This grid helps us visualize and locate points in the plane and is central to understanding the impact of a linear transformation.

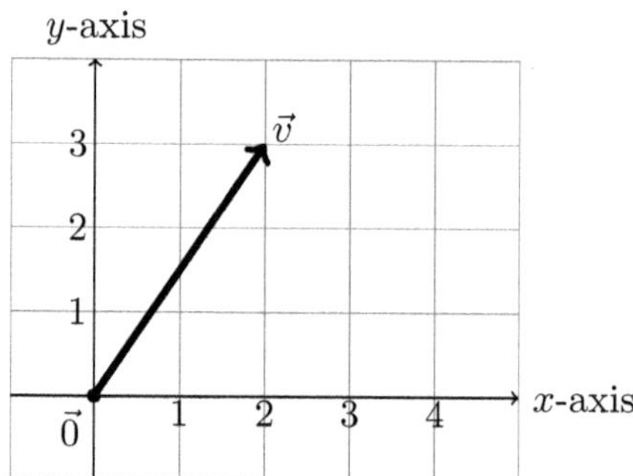

We will use a sample vector $\vec{v}$ to demonstrate some fundamental properties of vectors. If we assume that the grid lines are spaced one unit apart, then the vector $\vec{v}$ above begins at the origin and terminates at point (2,3) and is represented as:

$$\vec{v} = \begin{bmatrix} 2 \\ 3 \end{bmatrix}$$

The direction of the arrow indicates the direction of $\vec{v}$, while the length of the arrow represents its *magnitude* or *length*. Recall that the formula for distance d between two points (x_1, y_1) and (x_2, y_2) in the coordinate plane follows easily from the Pythagorean theorem is given by:

$$d = \sqrt{(x_2 - x_1)^2 + (y_2 - y_1)^2}$$

For a vector $\vec{v}$, its length or magnitude, denoted by absolute value, is calculated as the distance from the origin to its endpoint: given by $|\vec{v}| = \sqrt{v_x^2 + v_y^2}$. For our example, we find: $|\vec{v}| = \sqrt{2^2 + 3^2} = \sqrt{13}$

Next, let us add another vector to our coordinate plane. We now have two vectors: $\vec{v} = \begin{bmatrix} 2 \\ 3 \end{bmatrix}$ and $\vec{u} = \begin{bmatrix} 4 \\ -1 \end{bmatrix}$. To visualize $\vec{u}$, draw an arrow starting from the origin $(0, 0)$ to the point $(4, -1)$ on a coordinate plane as shown below.

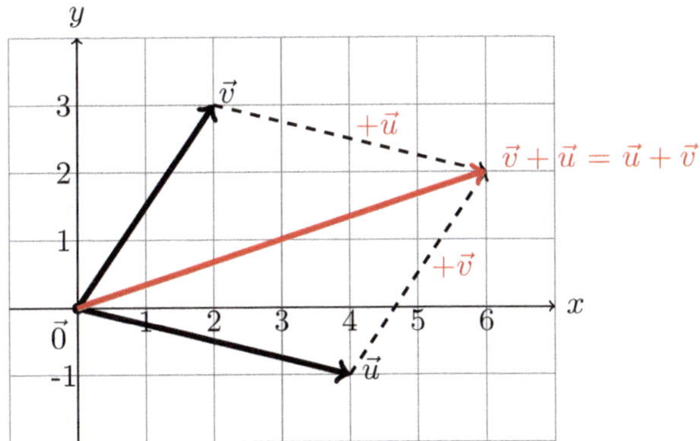

Exercise 6.3 Compute the length of $\vec{u}$.

A vector can be interpreted in two different ways: as a simple point in the Euclidean plane or, as the arrow indicates, the instruction to translate any point in the plane by a specified amount. In the figure, we attached a parallel copy of the vector $\vec{u}$ to the point of $\vec{v}$ and translate it to obtain the point labeled $\vec{v} + \vec{u}$. Translating the point $\vec{u}$ by the vector $\vec{v}$ giving $\vec{u} + \vec{v}$ and also completing a parallelogram giving, $\vec{u} + \vec{v} = \vec{v} + \vec{u}$.

This is a natural place to summarize the basic algebraic operations on vectors. There are two fundamental operations: vector addition and scalar multiplication. We have just introduced addition of vectors, $\vec{v} + \vec{u}$, geometrically. From this geometric definition, we can deduce that if

$$\vec{v} = \begin{bmatrix} v_x \\ v_y \end{bmatrix} \quad \text{and} \quad \vec{u} = \begin{bmatrix} u_x \\ u_y \end{bmatrix},$$

then

$$\vec{v} + \vec{u} = \begin{bmatrix} v_x + u_x \\ v_y + u_y \end{bmatrix}.$$

From this, we can easily verify the basic rules of addition:

Rule 1. *Commutativity:* $\vec{u} + \vec{v} = \vec{v} + \vec{u}$, for all vectors $\vec{u}$ and $\vec{v}$.

Rule 2. *Associativity:* $(\vec{u} + \vec{v}) + \vec{w} = \vec{u} + (\vec{v} + \vec{w})$, for all vectors $\vec{u}, \vec{v}, \vec{w}$.

Rule 3. *Additive identity:* There exists a vector $\vec{0} = \begin{bmatrix} 0 \\ 0 \end{bmatrix}$ such that $\vec{0} + \vec{v} = \vec{v}$, for all vectors $\vec{v}$.

Rule 4. *Additive inverse:* Every vector $\vec{v}$ has an inverse $-\vec{v} = \begin{bmatrix} -v_x \\ -v_y \end{bmatrix}$ such that $-\vec{v} + \vec{v} = \vec{0}$.

In addition to vector addition, there is another algebraic operation: *scalar multiplication.* If s is a scalar (a real number) and $\vec{v} = \begin{bmatrix} v_x \\ v_y \end{bmatrix}$ is a vector, then

$$s\vec{v} = \begin{bmatrix} s v_x \\ s v_y \end{bmatrix}.$$

The rules for scalar multiplication are:

Rule 5. $1\vec{v} = \vec{v}, \quad 0\vec{v} = \vec{0}, \quad$ and $\quad (-1)\vec{v} = -\vec{v}$, for all vectors $\vec{v}$.

Rule 6. *Associativity:* $(st)\vec{v} = s(t\vec{v})$, for all scalars s, t and all vectors $\vec{v}$.

Rule 7. *Distributivity:* $s(\vec{u} + \vec{v}) = s\vec{u} + s\vec{v}$ and $(s + t)\vec{v} = s\vec{v} + t\vec{v}$, for all scalars s, t and all vectors $\vec{u}, \vec{v}$.

Just as we often calculate the distance between two points in the coordinate plane, we frequently need to find the distance between the endpoints of two vectors, $\vec{v}$ and $\vec{u}$.

To do this, we simply apply the distance formula to their coordinate representations

$$|\vec{v} - \vec{u}| = \sqrt{(v_x - u_x)^2 + (v_y - u_y)^2}.$$

Exercise 6.4 Confirm that the distance between the points $\vec{v}$ and $\vec{u}$ in the example pictured above is $|\vec{v} - \vec{u}| = \sqrt{20}$.

To further our understanding of the Euclidean plane, we need to examine another fundamental geometric element: the line. The lines of the plane include two types: those that are the graphs of linear functions and vertical lines, which do not correspond to one-dimensional linear functions since their slope is undefined.

To make our discussions more consistent, it is helpful to have a uniform way of representing all lines in the plane. By using a parametric form, we can describe each line without needing a separate approach for vertical lines. This approach simplifies our analysis and is especially useful in proofs, where vertical lines would otherwise need to be treated as a unique case.

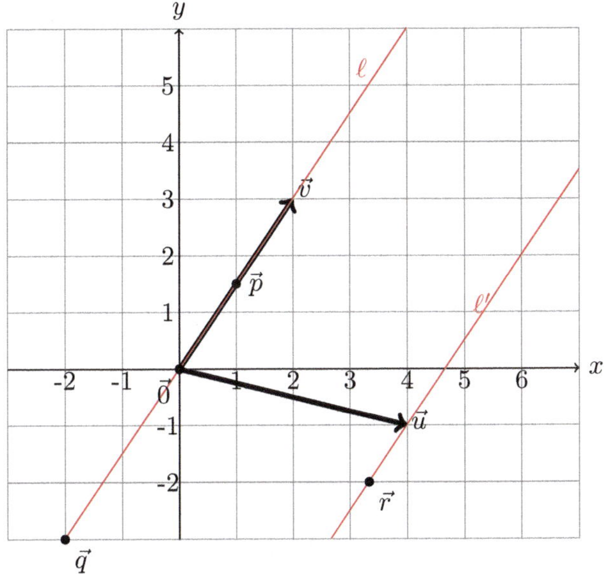

Lines through the origin have a simple parametric representation. For example, the line ℓ, pictured here, can be written parametrically as:

$$\ell = \{s\vec{v} \mid s \text{ is any real number}\},$$

where $\vec{v}$ is any fixed non-zero vector on ℓ, and s is the parameter. In this example, $\vec{v} = \begin{bmatrix} 2 \\ 3 \end{bmatrix}$ is the fixed non-zero vector, and the points $\vec{p}$ and $\vec{q}$ on ℓ are given by the parameters $\frac{1}{2}$ and -1, respectively:

$$\vec{p} = \begin{bmatrix} 1 \\ \frac{3}{2} \end{bmatrix} = \frac{1}{2}\begin{bmatrix} 2 \\ 3 \end{bmatrix} \qquad \vec{q} = \begin{bmatrix} -2 \\ -3 \end{bmatrix} = -1\begin{bmatrix} 2 \\ 3 \end{bmatrix}$$

Let's consider the parametric equation for a line ℓ' that is parallel to ℓ but does not pass through the origin. We begin with the parametric representation of ℓ, which passes through the origin, and then add a point, $\vec{u}$, on ℓ' to get the parametric equation for ℓ'.

$$\ell' = \{s\vec{v} + \vec{u} \mid s \text{ is any real number}\},$$

This method shows that any line can be described in this manner, with points on the line obtained by varying the parameter s. For instance, the point $\vec{r}$ on ℓ' can be calculated using the parameter value $-\frac{1}{2}$:

$$\vec{r} = \begin{bmatrix} 3 \\ -\frac{5}{2} \end{bmatrix} = \left(-\frac{1}{2}\right) \begin{bmatrix} 2 \\ 3 \end{bmatrix} + \begin{bmatrix} 4 \\ -1 \end{bmatrix}.$$

This looks very familiar! To exploit this familiarity, we make a few notational changes: let $\vec{v} = \vec{a}$, let $\vec{u} = \vec{b}$, let $s = x$, switch the order of the scalar and the vector, and write ℓ' as a function:

$$\ell'(x) = \vec{a}x + \vec{b}.$$

We may view this as a linear function from the real line into the coordinate plane, whose image is the line ℓ'. For example,

$$\ell'(0) = \vec{b} = \begin{bmatrix} 4 \\ -1 \end{bmatrix}, \qquad \ell'\left(-\frac{1}{2}\right) = \vec{r}.$$

Thus, we may interpret the parametric equation for ℓ' as a linear function, where $\vec{a}$ (or $\vec{v}$) plays the role of the slope of this linear function, and $\vec{b}$ (or $\vec{u}$) acts as the translation vector.

As mentioned before, the parametric form for lines permits us to describe all lines even vertical lines (that are not graphs of linear functions).

To describe the y-axis parametrically, we note that every point on the y-axis has an x-coordinate of 0. Thus, a parametric equation for the y-axis is:

$$s \begin{bmatrix} 0 \\ 1 \end{bmatrix},$$

where s is a parameter that can take any real value, allowing us to describe every point along the line.

Now, consider a line that is vertical but does not pass through the origin, intersecting the x-axis instead at some fixed coordinate t. Such a line can be described parametrically by starting with the vector

$$\begin{bmatrix} t \\ 0 \end{bmatrix},$$

which represents the intersection point of the line and the x-axis, and then adding multiples of a vertical displacement vector:

$$s \begin{bmatrix} 0 \\ 1 \end{bmatrix} + \begin{bmatrix} t \\ 0 \end{bmatrix}.$$

This equation simplifies to:

$$\left\{ \begin{bmatrix} t \\ s \end{bmatrix} \,\middle|\, s \in \mathbb{R} \right\}.$$

In this form, we see that t remains fixed while s varies freely, capturing the vertical movement along the line.

In our coordinate system, the vertical grid lines correspond to all such lines where t is an integer, and s is allowed to vary. Specifically, the points on these lines are represented as:

$$\begin{bmatrix} t \\ s \end{bmatrix}, \quad \text{where } t \in \mathbb{Z} \text{ and } s \in \mathbb{R}.$$

Similarly, horizontal grid lines are defined as lines where the y-coordinate remains constant at some integer s, while the x-coordinate t varies. These lines can be described by the set of points:

$$\begin{bmatrix} t \\ s \end{bmatrix}, \quad \text{where } s \in \mathbb{Z} \text{ and } t \in \mathbb{R}.$$

Together, the vertical and horizontal lines form the familiar *coordinate grid*, a structure that divides the plane into a network of squares. Each point on the plane can be uniquely identified by its position relative to these grid lines, making the coordinate grid a fundamental tool for geometric analysis.

To find the *slope* of lines ℓ and ℓ', we use the ratio $\frac{v_y}{v_x}$, giving a slope of $\frac{3}{2}$ for both example lines. Now, consider a linear function $f(x)$ with ℓ as its graph. Since ℓ passes through points $(0, 0)$ and $(2, 3)$, we substitute these into $f(x) = ax + b$, yielding

$$0 = f(0) = b$$

and

$$3 = f(2) = 2a + b.$$

Solving this system, we find

$$f(x) = \frac{3}{2}x.$$

For the line ℓ', which goes through points $\vec{u}$ and $v \stackrel{\rightarrow}{+} u$ with coordinates $(4, -1)$ and $(6, 2)$, respectively, we use the same approach. Substituting into $f(x) = ax + b$ yields

$$-1 = f(4) = 4a + b$$

and

$$2 = f(6) = 6a + b.$$

Solving this system gives

$$f(x) = \frac{3}{2}x - 7,$$

showing how the two lines are parallel but differ by a vertical shift.

Exercise 6.5 Consider the two points $\vec{p}, \vec{q}$ with coordinates (p_x, p_y) and (q_x, q_y), respectively. Compute the parametric formulation of ℓ, the line through these two points and, if $p_x \neq q_x$, the linear function with ℓ as its graph.

The slope of a vertical line is undefined, or equivalently, we may describe it as having an infinite slope. In the plane, two distinct lines are considered *parallel* if and only if they do not intersect. By convention, we also regard each line as parallel to itself. Hence, two lines in the plane are *not parallel* if and only if they intersect in exactly one point.

Lemma 6.1

(1) The slope of a line remains the same whether the line is expressed parametrically or as a linear function.
(2) Two lines are parallel if and only if they have the same slope.

Proof

(1) This was established by you in Exercise 6.5.
(2) To prove this, we rely on part (1), which guarantees that the slopes of lines are independent of the chosen representation (parametric or linear). Since the properties

of vectors have not yet been fully developed, we present the proof in terms of linear functions.

Let the equations of two lines be $f(x) = ax + b$ and $g(x) = a'x + b'$. Assume the two lines are not parallel, therefore they must intersect at some point. At the intersection, the y-values of the two lines are equal, so:

$$ax + b = a'x + b'.$$

Rearranging terms yields:

$$(a - a')x = b' - b.$$

Solving for x, we have:

$$x = \frac{b' - b}{a - a'},$$

which is a unique solution provided $a \neq a'$. Therefore, the lines intersect if and only if their slopes a and a' are different. Conversely, if $a = a'$, the lines are parallel and do not intersect. This completes the proof. $\square$

6.4   Visualizing 2D Linear Functions

Having established vectors as the building blocks of $\mathbb{R}^2$, the Euclidean plane, we now turn our focus to defining and visualizing the two-dimensional linear functions that map $\mathbb{R}^2$ onto $\mathbb{R}^2$. To do this, we mirror the structure of one-dimensional linear functions $f(x) = ax + b$, but with a and b represented by elements of $\mathbb{R}^2$, or vectors. These 2D linear functions allow us to explore how vectors can be shifted, rotated, reflected, or scaled, and how these operations impact the geometric shapes formed by collections of vectors. Just as one-dimensional functions shift and scale points on the number line, 2D linear functions alter vectors within the plane.

A *2D linear function* has the form

$$L(\vec{v}) = \mathbf{A}\vec{v} + \vec{w}$$

where $\mathbf{A}$ is a 2×2 non-singular, matrix over the reals, referred to as the *slope matrix*, and $\vec{w}$ is a vector called the *translation vector*. When there is no translation vector (i.e., when $L(\vec{v}) = \mathbf{A}\vec{v}$), we call the function a *linear transformation*. These linear transformations represent transformations of $\mathbb{R}^2$ viewed as a vector space.

Note: Readers unfamiliar with matrices or in need of a review should pause here, complete the matrix review in the Appendix, and then return to this section. In particular, it is important to understand that the non-singularity condition on the matrix $\mathbf{A}$ forces our 2D linear functions to be one-to-one and onto. Without this non-singularity condition these functions could end up mapping the entire plane onto a single line or even a single point. This is analogous to certain one-dimensional linear functions—specifically, constant functions with zero slope—that map the entire real line onto a single point. In terms of matrices, transformations that collapse the plane in this way are represented by *singular* matrices, which are characterized by a determinant of zero.

Before delving further into the formal theory of two-dimensional linear functions, let's first explore how they transform the plane in ways analogous to how linear functions transform the line. The key to understanding how 2D linear functions operate lies in the following theorem.

Theorem 6.1 *Non-singular two-dimensional linear functions map lines onto lines and parallel lines onto parallel lines. Moreover, if $L : \mathbb{R}^2 \to \mathbb{R}^2$ is such a function and ℓ is a line in $\mathbb{R}^2$, then the restriction of L to ℓ is itself a linear function from ℓ to $L(\ell)$. Finally, within any equivalence class of parallel lines, the transformation stretches or shrinks distances along the lines by the same constant factor.*

Proof Let $L(\vec{v}) = \mathbf{A}(\vec{v}) + \vec{w}$ be a linear function in $\mathbb{R}^2$, and let ℓ be a line given parametrically by $\ell(s) = s\vec{v} + \vec{u}$. The image of ℓ under L is

$$L(\ell(s)) = \mathbf{A}(s\vec{v} + \vec{u}) + \vec{w} = s\mathbf{A}(\vec{v}) + \mathbf{A}(\vec{u}) + \vec{w}.$$

This represents a line with direction vector $\mathbf{A}(\vec{v})$ and passing through the point $\mathbf{A}(\vec{u}) + \vec{w}$. Thus, $L(\ell)$ is a line.

Now, suppose ℓ' is a line parallel to ℓ. Then, ℓ' has the same direction vector $\vec{v}$, and under L, its direction vector becomes $\mathbf{A}(\vec{v})$. Since ℓ and ℓ' are mapped to lines with the same direction vector $\mathbf{A}(\vec{v})$, $L(\ell')$ is parallel to $L(\ell)$.

To examine scaling, let $\vec{p} = s_p\vec{v} + \vec{u}$ and $\vec{q} = s_q\vec{v} + \vec{u}$ be two points on ℓ. The distance between $\vec{p}$ and $\vec{q}$ is

$$|\vec{p} - \vec{q}| = |(s_p\vec{v} + \vec{u}) - (s_q\vec{v} + \vec{u})| = |s_p - s_q| \cdot |\vec{v}|.$$

Under L, these points are mapped to

$$L(\vec{p}) = s_p\mathbf{A}(\vec{v}) + \mathbf{A}(\vec{u}) + \vec{w}, \quad L(\vec{q}) = s_q\mathbf{A}(\vec{v}) + \mathbf{A}(\vec{u}) + \vec{w}.$$

The distance between their images is

$$|L(\vec{p}) - L(\vec{q})| = |(s_p\mathbf{A}(\vec{v}) + \mathbf{A}(\vec{u}) + \vec{w}) - (s_q\mathbf{A}(\vec{v}) + \mathbf{A}(\vec{u}) + \vec{w})| = |s_p - s_q| \cdot |\mathbf{A}(\vec{v})|.$$

Thus, the distances are scaled by the factor

$$\frac{|\mathbf{A}(\vec{v})|}{|\vec{v}|}.$$

In other words, L restricted to ℓ is a similarity transformation, and thus by Theorem 5, a linear function. Finally, since all lines in a parallel class share the same direction vector $\vec{v}$, all distances along lines in that class are scaled by the same factor $\frac{|\mathbf{A}\vec{v}|}{|\vec{v}|}$. $\square$

Corollary 6.1 *Two-dimensional linear functions map the vertical grid lines onto a family of evenly spaced parallel lines and the horizontal grid lines onto a family of evenly spaced parallel lines.*

This corollary allows us to visualize the images of grid lines as two families of parallel lines, evenly spaced and forming a grid-like structure. Just as we represented the linear function f on a line ℓ by superimposing the image $f(\ell)$ onto the original scale of ℓ, we can represent a 2D linear function L by superimposing the transformed coordinate grid of $\mathbb{R}^2$ onto the original coordinate grid. This gives us a clearer picture of how the transformation alters the entire grid.

To illustrate, consider the 2D linear function $K = \mathbf{A}\vec{v} + \vec{w}$ where the slope matrix $\mathbf{A} = \begin{bmatrix} 1 & -1 \\ 1 & 1 \end{bmatrix}$ and the translation vector $\vec{w} = \begin{bmatrix} 1 \\ 1 \end{bmatrix}$. Thus, K acts on a vector $\vec{v} = \begin{bmatrix} v_x \\ v_y \end{bmatrix}$ as follows: $K(\vec{v}) = \begin{bmatrix} 1 & -1 \\ 1 & 1 \end{bmatrix} \begin{bmatrix} v_x \\ v_y \end{bmatrix} + \begin{bmatrix} 1 \\ 1 \end{bmatrix}$.

In Fig. 6.2 below, the standard coordinate grid is shown in black, with its image under K overlaid in red. The original coordinate axes and their transformed counterparts are drawn as thick lines.

Several observations can be made by examining this figure. First, the origin $\vec{0} = \begin{bmatrix} 0 \\ 0 \end{bmatrix}$ is mapped to $\begin{bmatrix} 1 \\ 1 \end{bmatrix}$, establishing this point as the origin of the red grid with coordinates $\begin{bmatrix} 0 \\ 0 \end{bmatrix}$ in the red grid. Next, we see that the unit vector along the x-axis, $\begin{bmatrix} 1 \\ 0 \end{bmatrix}$ (indicated by the blue point on the x-axis), is mapped to $\begin{bmatrix} 2 \\ 2 \end{bmatrix}$, which becomes the unit vector on the red x-axis (also marked by a blue point) with coordinates $\begin{bmatrix} 1 \\ 0 \end{bmatrix}$ in the red grid.. Similarly, the unit vector along the y-axis, $\begin{bmatrix} 0 \\ 1 \end{bmatrix}$ (represented by the orange point on the y-axis), is mapped to

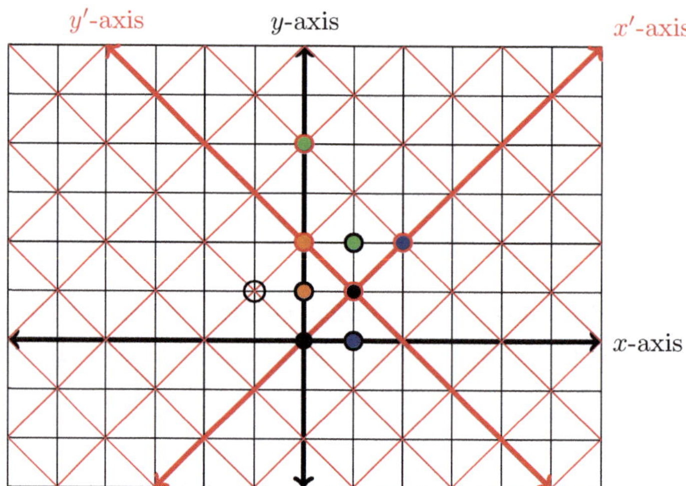

Fig. 6.2 Visualizing the 2D linear function K

$\begin{bmatrix} 0 \\ 2 \end{bmatrix}$, which corresponds to the orange point on the image of the y-axis and has coordinates $\begin{bmatrix} 0 \\ 1 \end{bmatrix}$ in the red grid. These values can be computed from the linear transformation formula by substitution.

$$K\begin{bmatrix} 0 \\ 0 \end{bmatrix} = \begin{bmatrix} 1 \\ 1 \end{bmatrix}, \quad K\begin{bmatrix} 1 \\ 0 \end{bmatrix} = \begin{bmatrix} 2 \\ 2 \end{bmatrix}, \quad K\begin{bmatrix} 0 \\ 1 \end{bmatrix} = \begin{bmatrix} 0 \\ 2 \end{bmatrix}.$$

To get a better feel for the transformation, we leave it as an exercise to find the image of some other points.

Exercise 6.6 Compute the following: $K\begin{bmatrix} -1 \\ 2 \end{bmatrix}$, $K\begin{bmatrix} 0 \\ 2 \end{bmatrix}$, $K\begin{bmatrix} 3 \\ -2 \end{bmatrix}$, $K\begin{bmatrix} 1 \\ 2 \end{bmatrix}$. Verify that in Fig. 6.2 the image of the green point, $\begin{bmatrix} 1 \\ 2 \end{bmatrix}$ under K is mapped to the green point $\begin{bmatrix} 0 \\ 4 \end{bmatrix}$ with coordinates $\begin{bmatrix} 1 \\ 2 \end{bmatrix}$ in the red grid. Check that the other three vectors above are mapped onto the point with the same coordinates in the red grid.

This visualization allows us to fully describe the linear transformation, K geometrically as follows: translate the entire grid by $\begin{bmatrix} 1 \\ 1 \end{bmatrix}$, taking the origin to $\begin{bmatrix} 1 \\ 1 \end{bmatrix}$; then rotate the grid $45°$ counterclockwise about $\begin{bmatrix} 1 \\ 1 \end{bmatrix}$, and stretch all lengths from $\begin{bmatrix} 1 \\ 1 \end{bmatrix}$ by a factor of $\sqrt{2}$.

While this construction allows us to find the images of individual points, it does not give a complete picture of how a two-dimensional linear transformation acts on the entire plane. Just as linear functions on the real line may have fixed points, a linear transformation in the plane may leave certain vectors unchanged. Identifying and understanding these fixed points is essential for analyzing the overall behavior of the transformation.

Given any two-dimensional linear function L and any point $\vec{c}$, we say that $\vec{c}$ is a *fixed point* of L, or that L *fixes* $\vec{c}$, if $L(\vec{c}) = \vec{c}$.

To determine whether the function K, shown above, has a fixed point

$$\vec{c} = \begin{bmatrix} c_x \\ c_y \end{bmatrix},$$

we solve the matrix equation

$$\begin{bmatrix} 1 & -1 \\ 1 & 1 \end{bmatrix} \begin{bmatrix} c_x \\ c_y \end{bmatrix} + \begin{bmatrix} 1 \\ 1 \end{bmatrix} = \begin{bmatrix} c_x \\ c_y \end{bmatrix}.$$

This equation is equivalent to the system $c_x - c_y + 1 = c_x$ and $c_x + c_y + 1 = c_y$, which yields the solution $c_x = -1$ and $c_y = 1$.

Exercise 6.7 In Fig. 6.2, verify that $\vec{c} = \begin{bmatrix} -1 \\ 1 \end{bmatrix} = \begin{bmatrix} -1 \\ 1 \end{bmatrix}$ is indeed the fixed point for the linear function K. Also verify that K can described concisely the 2D linear function K: that rotates the plane $45°$ counterclockwise about $\vec{c}$, and then stretch all lengths $\vec{c}$ by a factor of $\sqrt{2}$.

Exercise 6.8 Consider the following two-dimensional linear function

$$M(\vec{v}) = \begin{bmatrix} 2 & -1 \\ 1 & 2 \end{bmatrix} \begin{bmatrix} v_x \\ v_y \end{bmatrix} + \begin{bmatrix} -1 \\ 2 \end{bmatrix}.$$

a. Find the images of unit x vector and the unit y vector under M
b. Draw image of the standard coordinate grid after applying the function.
c. Find fixed point of M

d. Describe this 2D linear function.

Now consider $K : \mathbb{R}^2 \to \mathbb{R}^2$, $K(\vec{v}) = \mathbf{A}\vec{v}$, where the matrix $\mathbf{A}$ is

$$\mathbf{A} = \begin{bmatrix} \frac{3}{2} & 1 \\ \frac{1}{2} & 2 \end{bmatrix}.$$

We'll start by constructing the new coordinate grid under the 2D linear function K, in Figs. 6.3 and 6.4. In Fig. 6.3, we home in on the image of the standard unit square; we expand the construction to the entire grid in Fig 6.4. Since there is no translation vector, K leaves the origin fixed. As above, the end point of the unit vector $\vec{x}$ and its image are colored blue, and the unit vector $\vec{y}$ and its image are colored purple.

Specifically, the image of the unit square is transformed into a parallelogram with vertices: $K(\vec{0}) = \vec{0}$, $K(\vec{y}) = \begin{bmatrix} 1 \\ 2 \end{bmatrix}$, $K\begin{bmatrix} 1 \\ 1 \end{bmatrix} = \begin{bmatrix} \frac{5}{2} \\ \frac{5}{2} \end{bmatrix}$, $K(\vec{x}) = \begin{bmatrix} \frac{3}{2} \\ \frac{1}{2} \end{bmatrix}$. These four points define the vertices of the "unit square" in the image grid. Shaded in red, we see it is a parallelogram differing markedly from the standard unit square we are accustomed to. Once we have identified the unit square for a given 2D linear function, we can extend this to visualize the entire transformed grid. The following illustrates the transformed standard grid for the 2D linear function $K(\vec{v})$. In addition to the corners of the unit square, we also show the point $\begin{bmatrix} -2 \\ 3 \end{bmatrix}$ marked in green and it's image $K(\begin{bmatrix} -2 \\ 3 \end{bmatrix}) = \begin{bmatrix} 0 \\ 5 \end{bmatrix} = \begin{bmatrix} -2 \\ 3 \end{bmatrix}$

Exercise 6.9 Use the transformed coordinate grid to find the image of $\begin{bmatrix} -2 \\ 2 \end{bmatrix}$ under 2D linear function K.

Let's take this 2D linear function one step further and consider what happens when we reverse the horizontal and vertical components.

Fig. 6.3 A 2D linear function that distorts angle measure

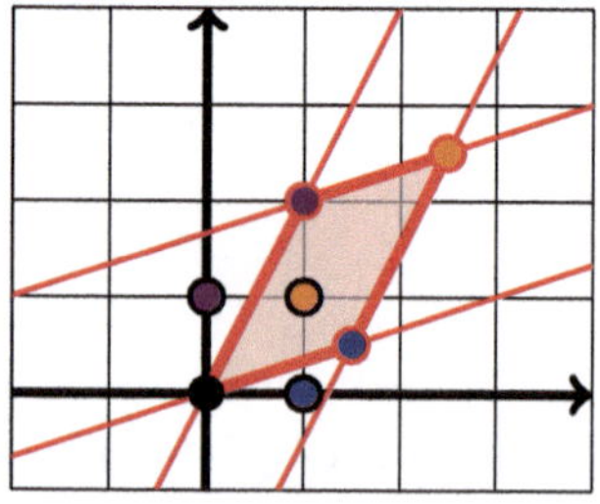

Fig. 6.4 The 2D linear
function K

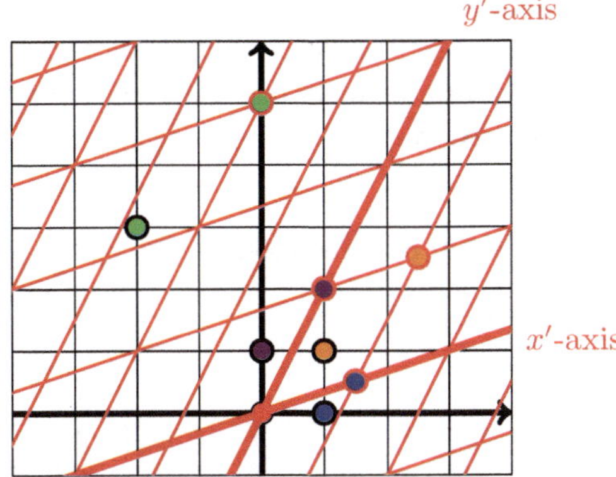

Fig. 6.5 The 2D linear
function H

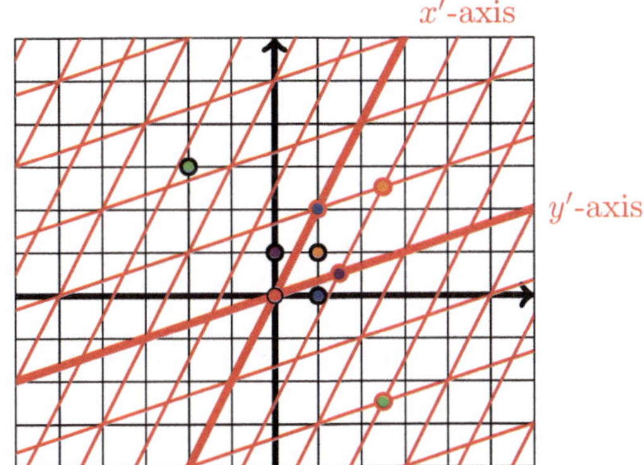

$$H(\vec{v}) = \begin{bmatrix} 1 & \frac{3}{2} \\ 2 & \frac{1}{2} \end{bmatrix} \begin{bmatrix} v_x \\ v_y \end{bmatrix}.$$

Here, the columns of the slope matrix are the same as for K, but in reverse order. This means that what was the red y axis in the image of K is now the x axis in the image of H, and the red y axis for K is now the red x axis for H (Fig. 6.5).

Once again, we can use the coordinate grids to identify the image of points under the transformation. For instance, the green point located at $\begin{bmatrix} -2 \\ 3 \end{bmatrix}$ on the standard black grid is mapped to the corresponding green point $\begin{bmatrix} -2 \\ 3 \end{bmatrix}$ on the red grid. However, it's important to observe that under transformations K and H, the point $\begin{bmatrix} -2 \\ 3 \end{bmatrix}$ is mapped to two distinct locations when viewed from the perspective of the standard grid.

Transformation K applies a non-uniform scaling to the axes, stretching them by different factors and bringing them closer together, similar to the effect of a retractable baby gate. Transformation H behaves similarly in that it stretches the axes as well, but it swaps the axes first.

Exercise 6.10 Find the image of $\vec{v} = \begin{bmatrix} -4 \\ 5 \end{bmatrix}$ and $\vec{u} = \begin{bmatrix} 4 \\ -1 \end{bmatrix}$ in the new coordinate grid shown below.

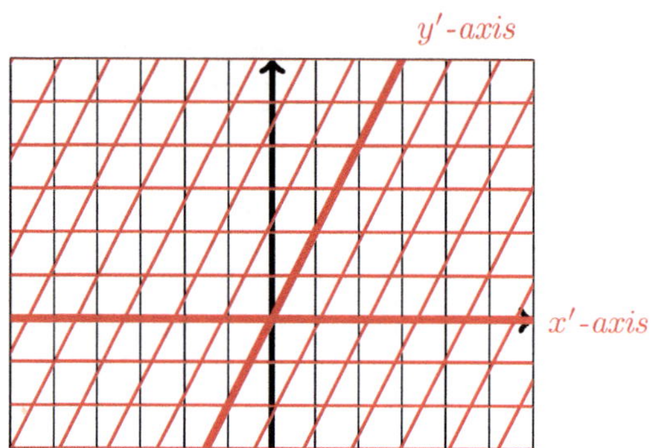

Exercise 6.11 Find the coordinates of the unit square for the linear transformation $G(\vec{v}) = \begin{bmatrix} 2 & \frac{1}{2} \\ \frac{3}{2} & 1 \end{bmatrix} \begin{bmatrix} v_x \\ v_y \end{bmatrix}$. Then extend this to depict the image of the standard coordinate grid under the transformation G.

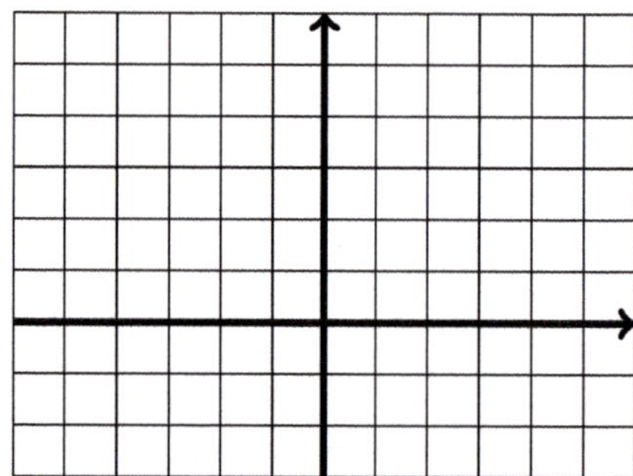

Before proceeding to the next section, we provide two examples of general 2D linear functions that collapse the plane onto either a single point or a line:

$$P(\vec{v}) = \begin{bmatrix} 0 & 0 \\ 0 & 0 \end{bmatrix} \begin{bmatrix} v_x \\ v_y \end{bmatrix} + \begin{bmatrix} 1 \\ 1 \end{bmatrix} \quad \text{and} \quad Q(\vec{v}) = \begin{bmatrix} 1 & -1 \\ 1 & -1 \end{bmatrix} \begin{bmatrix} v_x \\ v_y \end{bmatrix} + \begin{bmatrix} 0 \\ 0 \end{bmatrix}$$

The first function, P, maps the entire plane onto the point $\begin{bmatrix} 1 \\ 1 \end{bmatrix}$, while Q maps the plane onto the line given parametrically by $s \begin{bmatrix} 1 \\ 1 \end{bmatrix}$. Readers should verify that both matrices are singular.

6.5 Fixed Points and Lines

In the previous section, we explored the concept of fixed points for 2D linear functions. To recall, for a two-dimensional linear function L and a point $\vec{p}$, we say that $\vec{p}$ is a *fixed point* of L if $L(\vec{p}) = \vec{p}$. In other words, a fixed point is a point that remains unchanged under the transformation.

In addition to fixed points, a 2D linear function can also map a line onto itself. Specifically, for a line ℓ, we say that L *fixes* ℓ if $L(\ell) = \ell$. Furthermore, we say that L *fixes* ℓ *point-wise* if it fixes every point on ℓ, meaning that each point on the line remains unchanged under the transformation.

When a two-dimensional linear function has a unique fixed point, this point is known as the *center* of the function and plays a crucial role in understanding that function's behavior.

In Chap. 2, we established in Lemma 2.1 that, if two linear functions agree at two distinct points, they must be identical. In this chapter, we will extend this idea to linear transformations, demonstrating that a linear transformation is uniquely determined by its action on any three non-collinear points, which we will refer to as the Three Point Theorem for linear transformations.

Theorem 6.2 *If $\vec{p}, \vec{q},$ and $\vec{r}$ are three non-collinear points, and L and K are two linear transformations such that $L(\vec{p}) = K(\vec{p})$, $L(\vec{q}) = K(\vec{q})$, and $L(\vec{r}) = K(\vec{r})$, then $L = K$.*

Proof Let $\vec{p}' = L(\vec{p}) = K(\vec{p})$, $\vec{q}' = L(\vec{q}) = K(\vec{q})$, and $\vec{r}' = L(\vec{r}) = K(\vec{r})$. Let $\ell_{\vec{p}\vec{q}}$ be the line through $\vec{p}$ and $\vec{q}$. By Theorem 6.1, $L(\ell_{\vec{p},\vec{q}})$ and $K(\ell_{\vec{p},\vec{q}})$ are lines, and they intersect in two points, so they are the same line.

Furthermore, the restrictions of L and K to $\ell_{\vec{p},\vec{q}}$ are linear functions mapping $\ell_{\vec{p},\vec{q}}$ to $\ell'_{\vec{p},\vec{q}}$ that agree at two points. By Theorem 1, the Two Point Theorem, these functions must be identical, so L and K agree at all points on the line $\ell_{\vec{p},\vec{q}}$. By similar reasoning, L and K agree at all points on the lines $\ell_{\vec{p},\vec{r}}$ and $\ell_{\vec{q},\vec{r}}$.

Finally, let $\vec{z}$ be any point not on any of these three lines, and let ℓ' be any line containing $\vec{z}$ but none of the points $\vec{p}, \vec{q},$ or $\vec{r}$. Since ℓ' can be parallel to at most one of these lines, it must intersect at least two of them. Suppose that it intersects $\ell_{\vec{p},\vec{q}}$ at $\vec{q}'$ and $\ell_{\vec{p},\vec{r}}$ at $\vec{r}'$. Now, since L and K agree at both $\vec{q}'$ and $\vec{r}'$, they must also agree at all points on the line ℓ', including $\vec{z}$. Therefore, $L = K$. $\qquad\square$

We are now ready to explore the classification of fixed points and fixed lines for 2D linear functions. These geometric invariants play a key role in understanding the structure and behavior of linear transformations in the plane. This classification provides a systematic framework for determining the existence of fixed points and lines and reveals how their presence depends on the properties of the transformation matrix associated with L.

Theorem 6.3 *Consider the non-singular 2D linear function $L(\vec{v}) = \mathbf{A}\vec{v} + \vec{w}$.*

(1) L will have a unique fixed point if and only if $(\mathbf{I} - \mathbf{A})$ has an inverse (where $\mathbf{I}$ is the identity matrix) and that fixed point is $\vec{c} = (\mathbf{I} - \mathbf{A})^{-1}\vec{w}$. Furthermore, if L has a unique fixed point, we can write it in "slope-center" parallel to the 1-dimensional linear functions: $L(\vec{v}) = \mathbf{A}\vec{v} + (\mathbf{I} - \mathbf{A})\vec{c}$.

(2) If L has more than one fixed point, then either L is the identity (all points are fixed) or there is a line of fixed points ℓ, specifically, all points on ℓ are fixed and and no other points are fixed.

(3) If L has a fixed, but not point-wise fixed, line ℓ then either the restriction of L to ℓ is a translation or ℓ contains a fixed point c and the restriction L to ℓ is either a reflection with center c or a dilation with center c and multiplicative factor different from 1 and 0.

Proof

(1) Suppose that $\vec{c}$ is a fixed point of L. Then $L(\vec{c}) = \vec{c}$ or equivilantly $\mathbf{A}\vec{c} + \vec{w} = \vec{c}$. Rearranging terms gives:

$\mathbf{A}\vec{c} + \vec{w} = \mathbf{I}\vec{c}$ or $\vec{w} = (\mathbf{I} - \mathbf{A})\vec{c}$ and $L(\vec{v}) = \mathbf{A}\vec{v} + (\mathbf{I} - \mathbf{A})\vec{c}$.

This system has a unique solution if and only if $(\mathbf{I} - \mathbf{A})$ has an inverse. In this case, the solution is:

$$\vec{c} = (\mathbf{I} - \mathbf{A})^{-1}\vec{w}.$$

The matrix equation we are trying to solve for $\vec{c}$ is, $\vec{w} = (\mathbf{I} - \mathbf{A})\vec{c}$ is a system of two equations in two variables (the coordinates of $\vec{c}$). Thus it may fail to have a unique solution in two ways: either there is no solution (no fixed point) or there are many solutions (many fixed points).

(2) If there are two fixed points, say $L(\vec{p}) = \vec{p}$ and $L(\vec{q}) = \vec{q}$, let ℓ denote the line containing both $\vec{p}$ and $\vec{q}$. Then L maps ℓ onto itself. The restriction of L to ℓ is a linear function. Since this restriction has two fixed points, it must be the identity function on ℓ, fixing every point on ℓ.

If L fixes one additional point not on ℓ, then L would agree with the identity 2D linear function at three non-collinear points. By the previous theorem, L would then be the identity function on the entire plane.

(3) Let ℓ be a fixed line, but not a point-wise fixed line. Then the restriction of L to ℓ is a similarity transformation on ℓ but not the identity. The options are a translation or a reflection about a fixed point c on ℓ or a dilation with center c and multiplicative factor different from 1.

$\square$

Exercise 6.12   Given the 2D linear transformation L defined by the matrix

$$L(\vec{v}) = \begin{bmatrix} 2 & 1 \\ 0 & 1 \end{bmatrix} \vec{v},$$

find all the fixed points of L.

Exercise 6.13   Let $\vec{p}_1 = \begin{bmatrix} 1 \\ 2 \end{bmatrix}$, $\vec{p}_2 = \begin{bmatrix} 3 \\ 4 \end{bmatrix}$, and $\vec{p}_3 = \begin{bmatrix} 5 \\ 0 \end{bmatrix}$ be three non-collinear points in $\mathbb{R}^2$. Suppose $L : \mathbb{R}^2 \to \mathbb{R}^2$ is a 2D linear function such that:

$$L(\vec{p}_1) = \begin{bmatrix} 2 \\ 3 \end{bmatrix}, \quad L(\vec{p}_2) = \begin{bmatrix} 6 \\ 7 \end{bmatrix}, \quad \text{and} \quad L(\vec{p}_3) = \begin{bmatrix} 10 \\ 1 \end{bmatrix}.$$

a. Is the transformation L uniquely determined by its action on the points $\vec{p}_1$, $\vec{p}_2$, and $\vec{p}_3$? Explain your reasoning.
b. Given that the standard basis vectors in $\mathbb{R}^2$ are $\vec{e}_1 = (1, 0)$ and $\vec{e}_2 = (0, 1)$, express $L(\vec{e}_1)$ and $L(\vec{e}_2)$ as linear combinations of $L(\vec{p}_1)$, $L(\vec{p}_2)$, and $L(\vec{p}_3)$.
c. Use the results from part (b) to find the formula for L.

Applications

7

In this chapter, we shift our focus to real-world applications of two-dimensional linear functions, extending the one-dimensional cases developed earlier in the book. Linear transformations offer a powerful framework for modeling relationships and change in a wide range of settings, from financial planning to environmental systems. By exploring these applications, we will see how linear functions illuminate practical problems, uncovering patterns that are both mathematically elegant and deeply relevant to real-world decision-making.

7.1 Managing Student Loans and Savings

Many financial applications in one dimension, such as tracking a single savings account or managing a loan repayment, can be made more complicated by incorporating multiple cash streams. We'll extend our financial application from Chap. 3 by introducing two streams of cash flow such as simultaneously paying off a loan while building savings or managing multiple accounts with different interest rates and restrictions.

Matrices can model changes in multiple accounts over time. Consider a student, Bethyn, who graduates with a \$30,000 student loan, secures a job paying \$84,000 annually, and wishes to plan a budget to pay off the loan in 5 years while simultaneously building a savings account. The loan carries a 5% interest rate, compounded monthly, and the savings account offers a 4% interest rate, compounded monthly. Linear transformations allow Bethyn to track how much remains owed on the loan and how much she has available in her savings account over time.

Since the loan payments are due monthly, we are more concerned with Bethyn's monthly income rather than her annual income so we divide her annual income by 12, $\frac{84,000}{12} = 7000$. Bethyn decides that taking into account taxes, rent, groceries and other expenses she can comfortably live on 80% of her income, leaving the remaining 20% to put towards her financial goals–both loan repayment and savings. We use her monthly income to calculate her allocation for loan repayment and savings: $0.20 \times 7000 = \$1400$

Initially Bethyn assumes she can pay off her \$30,000 loan in 5 years by paying \$1000 a month for 30 months. To see if her assumption is correct, we model this budget allocation using a linear transformation in $\mathbb{R}^2$, where the first component represents loan repayment and the second represents savings. Vector $\vec{b}$ represents the allocation of the \$1400 toward both the loan and the savings account.

$$\vec{b} = \begin{bmatrix} \text{Savings} \\ \text{Loan Repayment} \end{bmatrix} = \begin{bmatrix} 400 \\ 1000 \end{bmatrix}$$

The linear transformation that models Bethyn's financial situation is given by $\mathbf{A}\vec{x} + \vec{b}$, where $\mathbf{A}$ represents the growth factors due to interest rates ($1.00\overline{3}$ for the savings account since 4% annual interest corresponds $0.\overline{33}\%$ monthly interest and 5% corresponds to growth of $1.0041\overline{6}$ for the loan), and $\vec{b}$ represents the monthly contributions (\$400 to savings and \$1000 towards the loan).

$$\mathbf{A} = \begin{bmatrix} 1.00\overline{3} & 0 \\ 0 & 1.0041\overline{6} \end{bmatrix} \quad \text{and} \quad \vec{b} = \begin{bmatrix} 400 \\ 1000 \end{bmatrix}$$

We can also write the linear transformation in slope-center form as $\mathbf{A}\vec{x} + (\mathbf{I} - \mathbf{A})\vec{c}$ where $\vec{c}$ is the center of the linear transformation. We use our values above to find $\vec{c}$

$$\vec{c} = (\mathbf{I} - \mathbf{A})^{-1}\vec{b} = \begin{bmatrix} -0.00\overline{3} & 0 \\ 0 & -0.0041\overline{6} \end{bmatrix}^{-1} \cdot \begin{bmatrix} 400 \\ 1000 \end{bmatrix} = \begin{bmatrix} -120,000 \\ -240,000 \end{bmatrix},$$

Notice that the entries on the main diagonal of $\mathbf{A}$ will always have the form $1 + r$, where r is the growth rate. In this example, all off-diagonal entries of $\mathbf{A}$ are zero because the account balances change independently of one another.

In situations where different groups influence each other, these off-diagonal entries will be nonzero. However, they will not include the leading 1, since they represent external influences rather than the growth of an existing account balance.

Recall earlier we used linear difference equations to track growth from month to month. We will use an extension of these linear difference equations to understand where Bethyn's funds are over time. The amounts in the savings and loan accounts after n months, denoted by $\vec{x}_n$, are given by the formula:

$$\vec{x}_n = \mathbf{A}^n(\vec{x}_0 - \vec{c}) + \vec{c},$$

Since we start with no savings and a principal balance of \$30,000 owed on Bethyn's loan, the initial state $\vec{x}_0$ is:

$$\vec{x}_0 = \begin{bmatrix} 0 \\ -30{,}000 \end{bmatrix}$$

Evaluating the linear difference equation for $n = 6$ months, we find:

$$\vec{x}_6 = \begin{bmatrix} 2420.09 \\ -24{,}695.01 \end{bmatrix}$$

This result shows that after six months, the savings account has accumulated \$2420.09 (including \$2400 in deposits and \$20.09 in interest), while the prinicipal loan balance has decreased to \$24,695.01.

For $n = 30$, we find $\vec{x}_{30} = \begin{bmatrix} 12{,}598.46 \\ -2100.61 \end{bmatrix}$. This shows that after diligently making payments for 5 years, Bethyn has nearly paid off her loan, but there's still a remaining balance. This highlights an error in her initial plan: paying \$1000 a month for 30 months didn't account for the interest accumulating on the loan.

Exercise 7.1 Determine how much Bethyn should pay each month if she wants to have the loan payed off in exactly 5 years. If she pays this amount each month, how much will she have accumulated in her savings account at the end of the 5 years?

Now let us consider another student, Ellen. Upon graduating—and without needing to take out a student loan—Ellen receives a generous \$15,000 gift from her Grandpa Joe to help her build financial stability. She decides to grow her savings by splitting the money between two account options: a high-yield account with withdrawal restrictions and a regular account that offers more flexibility. At the end of each month, the interest earned in the high-yield account is automatically transferred to the regular account. As a result, the matrix that represents this system becomes more intricate, since the transformation must account for how the balances shift between the two accounts.

In order to see how Ellen's gift will grow with time, we define a system where: Account 1 is a high-yield savings account earning 5% annual interest, compounded monthly and Account 2 is a regular savings account earning 2% annual interest, compounded monthly. At the end of each month, the interest earned in Account 1 is automatically transferred to Account 2 since there are no restrictions on withdrawals from Account 2. Our goal is to model how the balances in these accounts evolve over time using a two-dimensional linear transformation.

Ellen decides to put \$10,000 into the high yield account and keep \$5000 more accessible in the other account. We represent these initial values by the vector:

$$\vec{v}_0 = \begin{bmatrix} 10{,}000 \\ 5000 \end{bmatrix}$$

where the first entry corresponds to the balance in Account 1, and the second entry corresponds to the balance in Account 2.

Next, we define the monthly interest rates. For Account 1, the monthly interest rate is $\frac{5\%}{12} = 0.0041\overline{6}$, and for Account 2, it is $\frac{2\%}{12} = 0.001\overline{6}$. At the end of each month, the interest earned in Account 1 is transferred to Account 2. This interaction can be modeled by the following matrix:

$$\mathbf{A} = \begin{bmatrix} 1 & 0 \\ 0.0041\overline{6} & 1.001\overline{6} \end{bmatrix}$$

The first row of this matrix reflects that Account 1 retains its balance each month, while the second row reflects that interest from Account 1 is added to Account 2, in addition to the regular monthly interest on Account 2.

To update the balances after one month, we apply the matrix transformation:

$$\vec{v}_1 = \mathbf{A}\vec{v}_0 = \begin{bmatrix} 1 & 0 \\ 0.0041\overline{6} & 1.001\overline{6} \end{bmatrix} \begin{bmatrix} 10{,}000 \\ 5000 \end{bmatrix}$$

Carrying out the matrix multiplication, we obtain:

$$\vec{v}_1 = \begin{bmatrix} 10{,}000 \\ 0.0041\overline{6}(10{,}000) + 1.001\overline{6}(5000) \end{bmatrix} = \begin{bmatrix} 10{,}000 \\ 5050.005 \end{bmatrix}$$

Thus, after one month, the balance in Account 1 remains at $10{,}000$, while the balance in Account 2 increases to approximately $5{,}050.01$.

Applying the transformation again for the second month, we compute:

$$\vec{v}_2 = \mathbf{A}\vec{v}_1 = \begin{bmatrix} 1 & 0 \\ 0.0041\overline{6} & 1.001\overline{6} \end{bmatrix} \begin{bmatrix} 10{,}000 \\ 5050.005 \end{bmatrix}$$

This gives:

$$\vec{v}_2 = \begin{bmatrix} 10{,}000 \\ 0.0041\overline{6}(10{,}000) + 1.001\overline{6}(5050.005) \end{bmatrix} = \begin{bmatrix} 10{,}000 \\ 5100.09 \end{bmatrix}$$

After two months, the balance in Account 2 grows to approximately 5100.09.

This process can be repeated over several months by recursively applying the matrix transformation. After n months, the balances in both accounts are given by:

$$\vec{v}_n = \mathbf{A}^n \vec{v}_0$$

where $\mathbf{A}^n$ represents the matrix $\mathbf{A}$ raised to the power of n, which accounts for the compounding interest and transfers between accounts.

Exercise 7.2   What will be the balance in Account 2 after 5 years?

Exercise 7.3   How will this problem change if we start by putting the entire initial gift into the high-yield account or alternatively putting all funds into the regular savings account?

Throughout our lives, we face constant financial choices, whether it is budgeting, saving, or investing. These mathematical tools are valuable for anyone looking to gain financial confidence and make well-informed decisions. With these tools, we can effectively plan for major life events, such as retirement, by projecting how our savings and investments will grow over time. We can develop repayment strategies for loans, ensuring that we understand the long-term impact of interest rates and payment schedules. Additionally, we can analyze the benefits of different investment options, comparing how various rates of return will affect our financial future.

Exercise 7.4   Linnea wants to invest in a mutual fund that is expected to earn 5% annually and pay a 1% monthly dividend to a savings account that earns 3% interest. Dividends are transferred to the savings account each month, while the interest earned from the mutual fund is reinvested. Write a matrix A that models the interactions between the mutual fund and the savings account in this financial setup.

Exercise 7.5   Use matrix $\mathbf{A}$ from the previous exercise to determine how much Linnea will have in her account if she begins with no money in the mutual fund and $500 in her savings account. Assume that she chooses to live on half of her $60,000 take-home income, investing the remaining half into the mutual fund each month.

In this section, we modeled two streams of financial investment using a 2×2 matrix. This matrix captured how funds moved between two accounts and how interest accumulated over time. Naturally, we might wonder whether these techniques can be extended to three streams of cash flow. For instance, what if we introduce a mutual fund, a checking account, or an additional loan? In that case, a 3×3 matrix can represent the interactions among the three accounts. Each row of the matrix corresponds to one account, with the entries modeling both the interest applied and the movement of funds between accounts. For example, the first row might represent a mutual fund earning interest, the second a savings account that receives transfers from the fund, and the third a checking account where any remaining funds are deposited.

This setup allows us to track contributions, interest rates, and transfers between the accounts simultaneously. By applying the matrix transformation repeatedly, we can observe how the balances in each account evolve over time, and how changes in one account impact the others. By extending from a 2 × 2 matrix to a 3 × 3 matrix, students gain a more powerful framework for modeling and solving financial problems involving multiple accounts or investments. We leave it to the student to start considering how these tools would work with more than three streams of financial investment.

Exercise 7.6 Suppose Jeremy has two investment accounts and a loan. The first investment account earns 4% annual interest, compounded monthly, while the second earns 6% annually. Jeremy makes monthly payments toward the loan, which has an annual interest rate of 5%. Construct a 3 × 3 matrix that models the flow of funds between the two investment accounts and the loan, where part of the interest from the first account is used to pay off the loan, and the remaining funds are reinvested.

Exercise 7.7 Design a financial scenario involving three distinct accounts, such as investment accounts, savings accounts, or loans. Then, create a 3 × 3 matrix that models the flow of funds between Jeremy's accounts. Be sure to explain the specific types of accounts you chose and describe how money moves between them, including factors like interest rates, contributions, and transfers.

7.2   Predator-Prey Model

We will now return to our example from Chap. 3 regarding the pond stocked with fish and we will extend this to a two-dimensional scenario. Let's assume Colin and Elliott are designing a self sustaining pond where they can fish 30 largemouth bass each year. They will stock the farm with bluegills, minnows and largemouth bass. Bluegill reproduce to feed the bass and the minnows provide food for both the bluegill and the bass until the point where the bluegill have established themselves enough to support the bass.

The two friends use a predator-prey model to illustrate how linear transformations can represent the interactions between species in an ecosystem. The population dynamics of the species can be modeled by the following linear system:

$$\vec{x}_{t+1} = \mathbf{A}\vec{x}_t,$$

where $\vec{x}_t = \begin{bmatrix} M_t \\ B_t \\ L_t \end{bmatrix}$ represents the populations of minnows, bluegills and largemouth bass at time t respectively. $\mathbf{A}$ is a 3 × 3 transformation matrix that encodes the growth rates and interactions between the species due to both natural processes and predation. For this example, we measure time in months.

The minnows initially serve as the primary food source for both the bluegills and the largemouth bass but they are slow moving and easy prey. Their population will deplete rapidly over the first year. Once the population of minnows has been depleted the bluegills will feed on insects and small crustaceans. Once the bluegill population reaches maturity, they will reproduce at a rate sufficient to sustain the bass, creating a more stable ecosystem.

By applying the linear transformation at each step, we can compute the population vector $\vec{x}_t$ after t months, allowing us to track the changes in the ecosystem and understand the long-term dynamics of the species.

The transformation matrix captures the interaction rates between species. While fish don't reproduce at a constant rate throughout the year, Colin and Elliott make this simplification to consider the iterations monthly instead of yearly. Using this matrix, they apply linear difference equations to calculate the population vector, $\vec{x}_t$, at any given time t in months.

$$\vec{x}_{n+1} = \begin{bmatrix} 1.065 & -0.15 & -0.15 \\ 0.02 & 1.015 & -0.15 \\ 0.0015 & 0.002 & 1 \end{bmatrix} \vec{x}_n$$

This matrix captures the interactions among the three species. The minnow population grows by 6.5% each month through reproduction and 15% of the minnow population is consumed by both bluegill and bass. The bluegill population increases by 1.5% each month from reproduction, gains an additional 2% by feeding on minnows, but loses 15% due to predation by largemouth bass. Lastly, all the bass population growth is due to their consumption of minnows and bluegills. The bass population grows by 0.35% per month with 0.15% of their growth attributed to eating minnows and 0.2% to eating bluegill.

Colin and Elliott stock their pond with 1200 fathead minnows, 800 bluegill, and 60 largemouth bass. This initial population can be represented as the following vector $x_0 =$
$$\begin{bmatrix} 1200 \\ 800 \\ 60 \end{bmatrix}$$
To compute the population vector after $t = 5$ months, we first compute $\mathbf{A}^5$ and then multiply it by $\vec{x}_0$. This calculation yields

$$\vec{x}_5 = \mathbf{A}^5 \vec{x}_0 = \begin{bmatrix} 1.33 & -0.876 & -0.605 \\ 0.114 & 1.042 & -0.8 \\ 0.009 & 0.08 & 0.995 \end{bmatrix} \begin{bmatrix} 1200 \\ 800 \\ 60 \end{bmatrix} = \begin{bmatrix} 862.88 \\ 922 \\ 76.63 \end{bmatrix}.$$

This result shows that after 5 years, the fathead minnow population has decreased to 863, the bluegill population has increased to 922, and the largemouth bass population has grown slightly to approximately 76.63.

Exercise 7.8 Use the linear transformation above to find how many minnows remain in the pond at the end of the first year. How many of the other fish are in the pond?

By the end of the first year, Colin and Elliott have established a self-sustaining pond with mature bluegills and largemouth bass, ready for recreational fishing. They plan to catch 36 bass each year.

With the depletion of the minnows from the system, we can now model the ongoing populations of bluegill and largemouth bass using a 2×2 system that captures the interactions between these two species. In this updated model, the bluegill population reproduces at a rate that sustains both its own growth and the food supply for the bass. Meanwhile, the bass population increases as they feed on the bluegill but decreases due to fishing, with 36 bass being caught annually.

The aim is to maintain a stable ecosystem, where the bluegill population grows just enough to sustain the bass population, and the bass population remains balanced by the rate of fishing.

Let the population of bluegill at month n be B_n, and the population of largemouth bass at month n be L_n. The evolution of the populations month-to-month can be modeled by the following system:

$$\begin{bmatrix} B_{n+1} \\ L_{n+1} \end{bmatrix} = \begin{bmatrix} 1.02 & -0.03 \\ 0.05 & 1.00 \end{bmatrix} \begin{bmatrix} B_n \\ L_n \end{bmatrix} + \begin{bmatrix} 0 \\ -3 \end{bmatrix}$$

Each month, the bluegill population reproduces at a rate of 2%, while 3% of the bluegill population is consumed by the bass. The largemouth bass population increases by 5% with all their growth attributed to their consumption of bluegill, The term -3 in the transformation matrix reflects the monthly removal of bass through fishing, as they plan to catch 36 bass annually, which translates to three bass being removed each month. Since this is a fixed number, rather than a percentage of the remaining bass, we use a translation matrix to account for the constant rate of fishing.

Starting with an initial population of $B_0 = 1009$ bluegill and $L_0 = 96$ bass, we can use this matrix system to model the population changes over time, aiming to keep the ecosystem relatively stable.

This setup maintains a balanced ecosystem because the bluegill population grows slightly each month, offsetting the loss from bass consumption. Meanwhile, the bass population benefits from feeding on the bluegill, but fishing ensures their population remains under control.

Exercise 7.9 How many of each type of fish will you have in your pond after 5 years? 10 years? 50 years?

Exercise 7.10 How would this system change if you wanted to fish one bass per week instead of just 3 a month. How many months could you continue this fishing habit before you ran out Bass

Exercise 7.11 Bluegills are also a desirable fish to catch and eat. How would the system change if you didn't release the bluegills you catch back into the pond? Would the system still remain stable?

7.3 Markov Processes

The final application we will explore in this chapter is the study of systems that change from one state to another based on specific probabilities, a concept known as Markov processes. These processes can be modeled using a transition matrix, where each entry represents the probability of transitioning from one state to another. When we apply the linear transformation defined by the transition matrix to a vector representing the current state of the system (such as population levels), the result is a new vector showing the state of the system after one time step. By repeatedly applying the transition matrix, we can observe how the system evolves over time, providing valuable insights into its long-term behavior.

For example, consider a county where the population is divided between urban and rural areas. Each year, a portion of the urban population moves to rural areas, while the rest remains in the city. Similarly, some rural residents stay where they are, while others move to the city. This movement can be modeled using a Markov chain, represented by a 2×2 transition matrix. Each entry in the matrix corresponds to the probability of transitioning between urban and rural areas, capturing the yearly shifts in population distribution linear transformation matrix.

Let's use a linear transformation $\mathbf{P}$ to represent the probabilities of movement between the city dwellers and those that live in a more rural setting:

$$\mathbf{P} = \begin{bmatrix} p_{UU} & p_{UR} \\ p_{RU} & p_{RR} \end{bmatrix},$$

where p_{UU} is the probability that a urban dweller stays in the city, p_{UR} is the probability that a urban dweller moves to the countryside, p_{RU} is the probability that a rural dweller moves to the city, and p_{RR} is the probability that a rural dweller stays in the countryside. Suppose the probabilities are as follows: 70% of the city dwellers stay in the city each year, so $p_{UU} = 0.7$; 30% of the city dwellers move to the countryside, so $p_{UR} = 0.3$; 40% of the rural dwellers move to the city, so $p_{RU} = 0.4$; and 60% of the rural dwellers stay in the countryside, so $p_{RR} = 0.6$. Using these values our transition matrix $\mathbf{P}$ becomes:

$$\mathbf{P} = \begin{bmatrix} 0.7 & 0.3 \\ 0.4 & 0.6 \end{bmatrix}.$$

The initial population distribution is represented by the vector $\vec{x}_0 = \begin{bmatrix} x_{U0} & x_{R0} \end{bmatrix}$, where x_{U0} is the proportion of the population that live in the city and x_{R0} is the proportion who live rurally. For example, if initially 75% of the population is in the city and 25% is in the countryside, then $\vec{x}_0 = \begin{bmatrix} 0.75 & 0.25 \end{bmatrix}$.

The population distribution after one year, $\vec{x}_1$, can be determined by applying the transition matrix $\mathbf{P}$ to the initial distribution $\vec{x}_0$:

$$\vec{x}_1 = \vec{x}_0 \cdot \mathbf{P} = \begin{bmatrix} 0.75 & 0.25 \end{bmatrix} \cdot \begin{bmatrix} 0.7 & 0.3 \\ 0.4 & 0.6 \end{bmatrix} = \begin{bmatrix} 0.625 & 0.375 \end{bmatrix}.$$

So after just one year the percentage of the population living in the city has decreased to 62.5% and the percent of the population living rurally has grown to 37.5%. This may lead us to wonder if the population living in the city will continue to decrease and if so, how far will the percentage drop. To observe the effect of the Markov chain over multiple years, we apply the transition matrix to the initaial vector n times:

$$\vec{x}_n = \vec{x}_0 \cdot \mathbf{P}^n$$

Let's calculate the distribution for our society after 10 years and 100 years to see if we can sense the trend.

$$\vec{x}_{10} = \begin{bmatrix} 0.75 & 0.25 \end{bmatrix} \cdot \begin{bmatrix} 0.7 & 0.3 \\ 0.4 & 0.6 \end{bmatrix}^{10} = \begin{bmatrix} 0.57143 & 0.42857 \end{bmatrix}$$

$$\vec{x}_{100} = \begin{bmatrix} 0.75 & 0.25 \end{bmatrix} \cdot \begin{bmatrix} 0.7 & 0.3 \\ 0.4 & 0.6 \end{bmatrix}^{100} = \begin{bmatrix} 0.57143 & 0.42857 \end{bmatrix}$$

As n increases and the distribution stops changing, we say the system has reached a steady state. In this example, the distribution indeed stabilizes, as it remains consistent from year 10 through year 100. The steady-state solution indicates that approximately 57% of the population resides in the city, while around 43% live in rural areas.

Exercise 7.12 Confirm that this distribution will continue to hold steady by verifying it is unchanged after 5000 years.

If initially 50% of the population is in the city and 50% is in the countryside, then the population distribution after one year is given by

$$\begin{bmatrix} 0.5 & 0.5 \end{bmatrix} \cdot \vec{x}_1 = \mathbf{P} \cdot \vec{x}_0 = \begin{bmatrix} 0.7 & 0.3 \\ 0.4 & 0.6 \end{bmatrix} = \begin{bmatrix} 0.5 \\ 0.5 \end{bmatrix}.$$

Interestingly, in this case, the population distribution does not change after one year, indicating a steady state. However, this outcome is specific to this particular initial distribution and usually will not hold if we choose different parameters as we saw in the example above.

Exercise 7.13 Russell has two favorite coffee shops, Shop X and Shop Y, and he alternates between them based on his schedule and preferences. From past patterns, Russell knows if he goes to Shop X one week, there's a 20% chance that he will visit Shop Y the next week, and if he goes to Shop Y one week, there's a 10% chance that he will visit Shop X the next week.

Set up a Markov process to represent this situation using a 2×2 matrix, where each element represents the probability of visiting either shop in the next week. Calculate the distribution of Russell's visits after 5 weeks, starting with $\vec{x}_0 = \begin{bmatrix} 1 & 0 \end{bmatrix}$ (i.e., Russell starts by visiting Shop X).

The Geometry of the Euclidean Plane

8

8.1   The Coordinatization of the Euclidean Plane

The Euclidean plane is homogeneous, meaning that it looks the same at every point. In addition, the plane is equipped with standard notions of distance and angle, which allow us to measure lengths and determine angular relationships. Using these tools, we can construct a coordinate system.

First, choose a point $\vec{0}$ to serve as the origin and two orthogonal lines ℓ_x and ℓ_y to serve as the coordinate axes. Typically, ℓ_x, the x-axis, is taken to be horizontal, and ℓ_y, the y-axis, is taken to be vertical. Next, assign a linear scale to each axis: on both axes, the origin $\vec{0}$ is labeled 0; on the x-axis, the point one unit (with respect to the plane's notion of length) to the right of $\vec{0}$ is labeled 1; and on the y-axis, the point one unit above $\vec{0}$ is labeled 1.

In this way, we obtain a two-dimensional vector space over the real numbers with orthogonal basis vectors $\vec{x}$ and $\vec{y}$, which are unit vectors along the x- and y-axes, respectively. Every point $\vec{p}$ in the plane can be written uniquely as a linear combination of these basis vectors:

$$\vec{p} = a\vec{x} + b\vec{y},$$

where a and b are real numbers called the x- and y-coordinates of $\vec{p}$, respectively.

We typically represent a vector $\vec{p}$ as either a row or column vector,

$$\vec{p} = [a, b] \quad \text{or} \quad \vec{p} = \begin{bmatrix} a \\ b \end{bmatrix}.$$

In particular,

© The Author(s), under exclusive license to Springer Nature Switzerland AG 2026
J. J. Edmond, J. E. Graver, *The Linear Function and Euclidean Geometry*, Synthesis Lectures on Mathematics & Statistics, https://doi.org/10.1007/978-3-032-22712-6_8

$$\vec{x} = [1, 0] = \begin{bmatrix} 1 \\ 0 \end{bmatrix} \quad \text{and} \quad \vec{y} = [0, 1] = \begin{bmatrix} 0 \\ 1 \end{bmatrix}.$$

With this vector space structure in place, we can now recover the geometric structure of the plane. We begin by introducing an algebraic method for measuring distances.

8.2   The Euclidean Inner Product

The *Euclidean inner product*, also known as the *dot product*, of two vectors

$$\vec{v} = \begin{bmatrix} v_x \\ v_y \end{bmatrix} \quad \text{and} \quad \vec{w} = \begin{bmatrix} w_x \\ w_y \end{bmatrix}$$

is defined by:

$$\vec{v} \cdot \vec{w} = \begin{bmatrix} v_x \\ v_y \end{bmatrix} \cdot \begin{bmatrix} w_x \\ w_y \end{bmatrix} = v_x w_x + v_y w_y.$$

Understanding the inner product of vectors opens the door to a deeper understanding of Euclidean geometry using algebraic tools. The inner product provides a mathematical framework for examining how vectors interact, enabling us to define and compute essential geometric quantities such as length, distance, and angles between vectors.

The *length* of the vector $\vec{v}$ is denoted by $|\vec{v}|$ and defined by:

$$|\vec{v}| = \sqrt{\vec{v} \cdot \vec{v}} = \sqrt{v_x^2 + v_y^2}.$$

The *distance* between two points $\vec{v}$ and $\vec{w}$ is given by:

$$|\vec{v} - \vec{w}| = \sqrt{(\vec{v} - \vec{w}) \cdot (\vec{v} - \vec{w})} = \sqrt{(v_x - w_x)^2 + (v_y - w_y)^2}.$$

Finally, the *cosine* of the angle between two vectors is given by

$$\cos(\angle \vec{v}\vec{0}\vec{w}) = \frac{\vec{v} \cdot \vec{w}}{|\vec{v}|\,|\vec{w}|},$$

where $\vec{0}$ is the origin. Since the only angles whose cosine is 0 are odd multiples of 90°, we conclude that nonzero vectors $\vec{v}$ and $\vec{w}$ are *perpendicular* or *orthogonal* ($\vec{v} \perp \vec{w}$) if and only if

$$\vec{v} \cdot \vec{w} = 0.$$

On the unit circle, the two standard definitions of cosine agree naturally. From linear algebra, the inner product of two vectors $\vec{u}$ and $\vec{v}$ is related to the angle θ between them by

$$\cos\theta = \frac{\vec{u}\cdot\vec{v}}{|\vec{u}|\,|\vec{v}|}.$$

If we take

$$\vec{u} = \begin{bmatrix} 1 \\ 0 \end{bmatrix} \qquad \text{and} \qquad \vec{v} = \begin{bmatrix} \cos\theta \\ \sin\theta \end{bmatrix}$$

on the unit circle, then $|\vec{u}| = |\vec{v}| = 1$. It follows that

$$\vec{u}\cdot\vec{v} = 1\cdot\cos\theta + 0\cdot\sin\theta = \cos\theta.$$

This coincides with the classical right–triangle definition, since on the unit circle the hypotenuse has length 1, the side adjacent to the angle θ has length $\cos\theta$, and therefore

$$\cos\theta = \frac{\text{adjacent}}{\text{hypotenuse}} = \frac{\cos\theta}{1} = \cos\theta$$

Thus the dot-product interpretation and the right–triangle interpretation are fully consistent, with the unit circle providing a natural bridge between them.

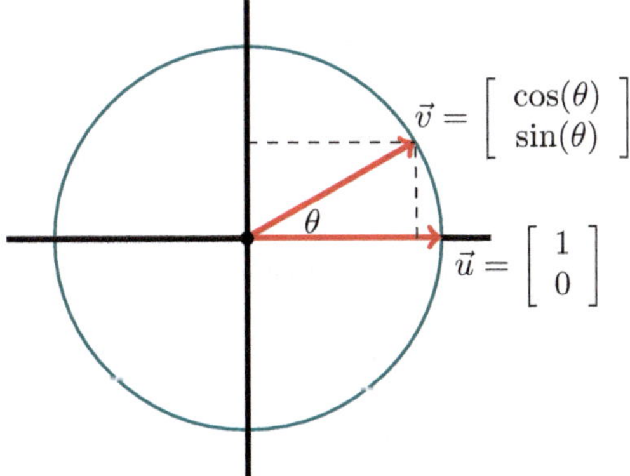

To justify the more general inner product definition of cosine, we begin with the unit circle centered at the origin, pictured on the left below. The standard coordinates for the unit vector that makes a counterclockwise angle of θ with the positive x-axis are

$$\begin{bmatrix} \cos(\theta) \\ \sin(\theta) \end{bmatrix}.$$

For the unit vector in the first quadrant, as shown above, these coordinates follow directly from the right triangle associated with the angle θ. Verifying that this coordinate representation remains valid in the second, third, and fourth quadrants requires the use of trigonometric identities, which we will examine later.

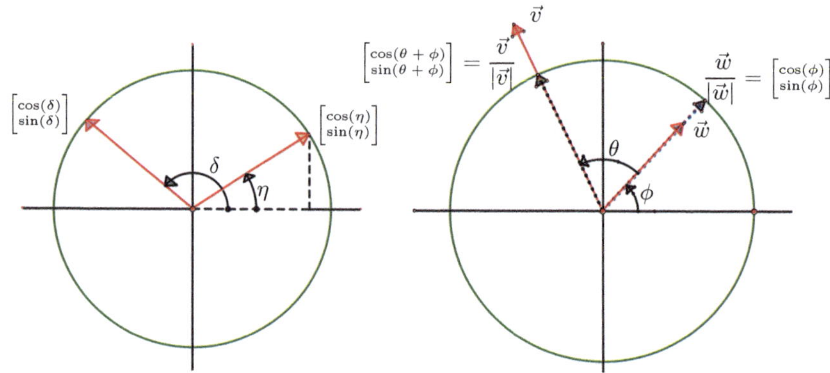

We now justify the formula used in the definition:

$$\cos(\angle \vec{v}0\vec{w}) = \frac{\vec{v}\cdot\vec{w}}{|\vec{v}||\vec{w}|} = \frac{\vec{v}}{|\vec{v}|}\cdot\frac{\vec{w}}{|\vec{w}|},$$

where $\frac{\vec{v}}{|\vec{v}|}$ and $\frac{\vec{w}}{|\vec{w}|}$ are the unit vectors in the directions of $\vec{v}$ and $\vec{w}$, respectively, as illustrated on the right in the figure. Thus, the cosine of the angle θ between $\vec{v}$ and $\vec{w}$ is defined as the inner product of the corresponding unit vectors:

$$\frac{\vec{v}}{|\vec{v}|}\cdot\frac{\vec{w}}{|\vec{w}|} = \begin{bmatrix}\cos(\theta+\phi)\\ \sin(\theta+\phi)\end{bmatrix}\cdot\begin{bmatrix}\cos(\phi)\\ \sin(\phi)\end{bmatrix} = \cos(\theta+\phi)\cos(\phi) + \sin(\theta+\phi)\sin(\phi).$$

To evaluate this inner product, we use the trigonometric identities

$$\cos(\theta+\phi) = \cos(\theta)\cos(\phi) - \sin(\theta)\sin(\phi),$$

$$\sin(\theta+\phi) = \sin(\theta)\cos(\phi) + \cos(\theta)\sin(\phi).$$

Substituting these into the inner product expression above gives

$$\big(\cos(\theta)\cos(\phi) - \sin(\theta)\sin(\phi)\big)\cos(\phi) + \big(\sin(\theta)\cos(\phi) + \cos(\theta)\sin(\phi)\big)\sin(\phi)$$

$$= \cos(\theta)\cos^2(\phi) - \sin(\theta)\sin(\phi)\cos(\phi) + \sin(\theta)\cos(\phi)\sin(\phi) + \cos(\theta)\sin^2(\phi).$$

The middle two terms cancel, leaving

$$\cos(\theta)\big(\cos^2(\phi) + \sin^2(\phi)\big) = \cos(\theta).$$

This confirms that the inner product definition of cosine agrees with its familiar geometric meaning.

Exercise 8.1 Consider the following points:

$$\vec{a} = \begin{bmatrix} 1 \\ 2 \end{bmatrix}, \quad \vec{b} = \begin{bmatrix} 4 \\ 1 \end{bmatrix}, \quad \vec{c} = \begin{bmatrix} 2 \\ -1 \end{bmatrix}.$$

a. Two of these vectors are orthogonal. Which two?
b. Which two points are closest together?
c. Find $\cos(\angle \vec{a}O\vec{b})$.

Having reviewed several geometric constructions, we can now use the inner product to express their algebraic counterparts.

- **Constructing a circle**
  Given a point $\vec{c}$ and a radius r, we use a pair of compasses to construct the circle with center $\vec{c}$ and radius r. Algebraically, the circle is defined by $|\vec{p} - \vec{c}| = r$, which gives the coordinates of the points $\vec{p}$ on the circle.
  For example, let $\vec{c} = \begin{bmatrix} 2 \\ -1 \end{bmatrix}$, $r = 5$, $\vec{p} = \begin{bmatrix} x \\ y \end{bmatrix}$. Then $\sqrt{(x-2)^2 + (y+1)^2} = 5$
  or $(x-2)^2 + (y+1)^2 = 25$. Some points with "nice" coordinates on this circle are shown in Fig. 8.1 below
- **Perpendicular bisector of a segment**
  For the segment $\overline{pq}$, the midpoint is

$$\vec{m} = \tfrac{1}{2}(\vec{p} + \vec{q}),$$

  The direction vector of the segment is

$$\vec{d}_s = \vec{p} - \vec{q}.$$

Exercise 8.2 Justify the midpoint and direction vector formulas given above.

To obtain the perpendicular bisector, we construct a direction vector $\vec{d}_b$ such that $\vec{d}_b \perp \vec{d}_s$. Then the line given parametrically by

Fig. 8.1 Circle constructions

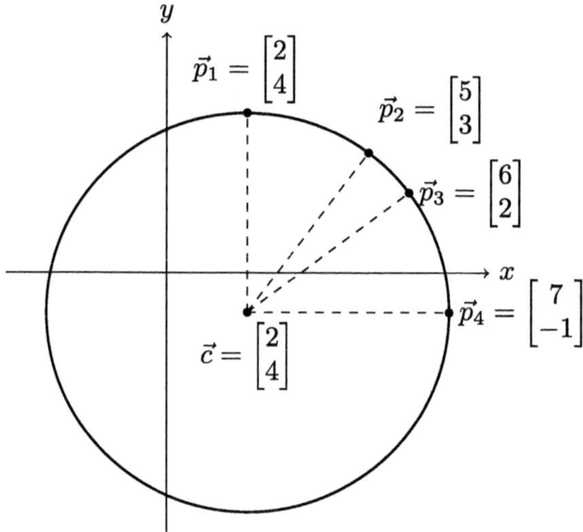

$$\ell(\vec{v}) = \lambda \vec{d}_b + \vec{m}$$

has its direction vector perpendicular to the segment and passes through the midpoint of that segment.

- **The circle containing three non-collinear points**
  Given three non-collinear points, there is a unique circle passing through all three. We can combine the algebraic constructions above (finding perpendicular bisectors, then their intersection) to determine the center and equation of this circle.

Exercise 8.3 Fill in the details of this construction. Then use it to verify that $\vec{c} = \begin{bmatrix} 2 \\ 4 \end{bmatrix}$ is the center of the center of the circle through the points $\vec{p}_1 = \begin{bmatrix} 2 \\ 4 \end{bmatrix}$, $\vec{p}_2 = \begin{bmatrix} 5 \\ 3 \end{bmatrix}$ and $\vec{p}_4 = \begin{bmatrix} 7 \\ -1 \end{bmatrix}$

Exercise 8.4 Find the circle passing through the following three points:

$$\vec{p}_1 = \begin{bmatrix} 1 \\ 2 \end{bmatrix}, \quad \vec{p}_2 = \begin{bmatrix} 4 \\ 3 \end{bmatrix}, \quad \vec{p}_3 = \begin{bmatrix} 2 \\ 5 \end{bmatrix}.$$

- **Tangents to a circle at a point on the circle.**

Given a circle C and a point $\vec{p}$ on C, the tangent line to C at $\vec{p}$ is constructed as the line through $\vec{p}$ perpendicular to the line joining $\vec{p}$ to the center of C.

Exercise 8.5 Find equations for the two tangent lines at $\vec{p}_1$ and $\vec{p}_3$ on the circle in Fig. 8.1.

In the examples above, we used algebra to describe classical geometric constructions such as circles, perpendicular bisectors, and tangent lines. Each of these constructions relied on the inner product to translate a geometric condition into an algebraic equation.

When we turn to triangles, we often wish to compute side lengths or angles given only partial information. For example, suppose we know the lengths of two sides of a triangle and the angle between them—how can we determine the length of the third side? Conversely, if all three sides are known, how can we find the measure of an angle?

To answer such questions, we require a general relationship between the sides and angles of a triangle. One very important tool for this purpose is the *Law of Cosines*.

Theorem 8.1 *Let $\vec{a}$, $\vec{b}$, and $\vec{c}$ be the vertices of a triangle, then*

$$|\vec{a} - \vec{c}|^2 = |\vec{a} - \vec{b}|^2 + |\vec{c} - \vec{b}|^2 - 2|\vec{a} - \vec{b}||\vec{c} - \vec{b}|\cos(\angle a\vec{b}c),$$

Proof We begin by verifying the vector identity

$$|\vec{a} - \vec{c}|^2 = |\vec{a} - \vec{b}|^2 + |\vec{c} - \vec{b}|^2 - 2(\vec{a} - \vec{b}) \cdot (\vec{c} - \vec{b}).$$

Expanding the left-hand side:

$$|\vec{a} - \vec{c}|^2 = (\vec{a} - \vec{c}) \cdot (\vec{a} - \vec{c})$$
$$= \vec{a} \cdot \vec{a} - 2\vec{a} \cdot \vec{c} + \vec{c} \cdot \vec{c}.$$

Expanding the right-hand side:

$$|\vec{a} - \vec{b}|^2 + |\vec{c} - \vec{b}|^2 - 2(\vec{a} - \vec{b}) \cdot (\vec{c} - \vec{b})$$
$$= (\vec{a} - \vec{b}) \cdot (\vec{a} - \vec{b}) + (\vec{c} - \vec{b}) \cdot (\vec{c} - \vec{b}) - 2(\vec{a} - \vec{b}) \cdot (\vec{c} - \vec{b})$$
$$= (\vec{a} \cdot \vec{a} - 2\vec{a} \cdot \vec{b} + \vec{b} \cdot \vec{b}) + (\vec{c} \cdot \vec{c} - 2\vec{c} \cdot \vec{b} + \vec{b} \cdot \vec{b}) - 2(\vec{a} \cdot \vec{c} - \vec{a} \cdot \vec{b} - \vec{b} \cdot \vec{c} + \vec{b} \cdot \vec{b})$$
$$= \vec{a} \cdot \vec{a} - 2\vec{a} \cdot \vec{c} + \vec{c} \cdot \vec{c}.$$

Here the $\vec{a} \cdot \vec{b}$, $\vec{c} \cdot \vec{b}$, and $\vec{b} \cdot \vec{b}$ terms cancel, confirming the identity.

What remains, is to verify that

$$2(\vec{a} - \vec{b}) \cdot (\vec{c} - \vec{b}) = 2|\vec{a} - \vec{b}| \, |\vec{c} - \vec{b}|\cos(\angle a\vec{b}c).$$

If we take $\vec{b}$ to be the origin, this simplifies to

$$2\,\vec{a}\cdot\vec{c} = 2|\vec{a}|\,|\vec{c}|\cos(\angle\vec{a}\,\vec{0}\,\vec{c}).$$

Dividing through by $2|\vec{a}||\vec{c}|$ gives

$$\cos(\angle\vec{a}\,\vec{0}\,\vec{c}) = \frac{\vec{a}\cdot\vec{c}}{|\vec{a}|\,|\vec{c}|},$$

which is the standard inner-product formula for the cosine of an angle. $\qquad\square$

Exercise 8.6 Consider a triangle with the following vertices

$$\vec{a} = \begin{bmatrix} 3 \\ 4 \end{bmatrix} \quad \vec{b} = \begin{bmatrix} 1 \\ 2 \end{bmatrix} \quad \vec{c} = \begin{bmatrix} 5 \\ 1 \end{bmatrix},$$

Use the Law of Cosines to determine whether this triangle is acute, right, or obtuse.

Exercise 8.7 Given two vectors

$$\vec{v} = \begin{bmatrix} 4 \\ 1 \end{bmatrix} \quad \vec{w} = \begin{bmatrix} 3 \\ 2 \end{bmatrix},$$

find a vector $\vec{u}$ such that the angle between $\vec{u}$ and $\vec{v}$ is $\frac{\pi}{3}$ and the angle between $\vec{u}$ and $\vec{w}$ is $\frac{\pi}{4}$.

Exercise 8.8 Consider two vectors $\vec{v}$ and $\vec{w}$, and let $\vec{t}$ be a translation vector.

$$\vec{v} = \begin{bmatrix} 2 \\ 3 \end{bmatrix}, \quad \vec{w} = \begin{bmatrix} 5 \\ 7 \end{bmatrix}, \quad \text{and} \quad \vec{t} = \begin{bmatrix} 1 \\ -2 \end{bmatrix}.$$

Show that the length $|\vec{v} - \vec{w}|$ is the same as the length of the vector difference after translation, $|(\vec{v} + \vec{t}) - (\vec{w} + \vec{t})|$.

Law of Cosines will be a useful tool in furthering our understanding of two-dimensional linear functions.

As established in Theorem 6.1, two-dimensional linear functions preserve key geometric structures. Specifically, they map points to points, lines to lines, and parallel lines to parallel lines. Moreover, the theorem shows that if a linear transformation L is restricted to a line ℓ, then it acts as a linear transformation from that line to its image $\ell' = L(\ell)$. Consequently, for any segment $\overline{pq}$ on ℓ, the length of its image $\overline{p'q'}$ on ℓ' is scaled by a constant factor, uniform across all lines parallel to ℓ. Our earlier work revealed that these scaling factors—often called *stretch factors*—can vary depending on direction.

The class of transformations that preserve both distances and angles are called *isometries* or *congruences*. Euclid described congruent figures intuitively: two figures are congruent "if one can be picked up and placed on top of the other, matching perfectly." We define two plane figures to be congruent if there exists an isometry of the plane mapping one onto the other.

Again referring to Theorem 6.1, we have that nonsingular 2D linear functions scale lengths along lines, that is, on each individual line they act as *similarities*. When such a transformation scales all lengths by the same factor, independent of direction, it is called a *similarity* of the Euclidean plane. In this framework, isometries are precisely the similarities with scale factor 1. Moreover, similarities preserve not only relative length but also angle measure.

Theorem 8.2 *Similarities preserve angle measure.*

Proof Let L be a similarity that alters lengths by the factor s, and consider $\angle a\vec{b}\vec{c}$, the angle at vertex $\vec{b}$ of the triangle $\triangle a\vec{b}\vec{c}$. Under L, the triangle $\triangle a\vec{b}\vec{c}$ is mapped to $\triangle L(\vec{a})L(\vec{b})L(\vec{c})$. Additionally, we know the following relations:

$$|L(\vec{a}) - L(\vec{b})| = s|\vec{a} - \vec{b}|, \quad |L(\vec{a}) - L(\vec{c})| = s|\vec{a} - \vec{c}|, \quad |L(\vec{c}) - L(\vec{b})| = s|\vec{c} - \vec{b}|.$$

Using the law of cosines for $\triangle L(\vec{a})L(\vec{b})L(\vec{c})$, we have:

$$|L(\vec{a})-L(\vec{c})|^2 = |L(\vec{a})-L(\vec{b})|^2+|L(\vec{c})-L(\vec{b})|^2-2|L(\vec{a})-L(\vec{b})||L(\vec{c})-L(\vec{b})|\cos(\angle L(\vec{a})L(\vec{b})L(\vec{c})).$$

Substituting the scaled lengths we get:

$$s^2|\vec{a} - \vec{c}|^2 = s^2|\vec{a} - \vec{b}|^2 + s^2|\vec{c} - \vec{b}|^2 - 2s|\vec{a} - \vec{b}| \cdot s|\vec{c} - \vec{b}|\cos(\angle L(\vec{a})L(\vec{b})L(\vec{c})).$$

Dividing through by s^2, yields:

$$|\vec{a} - \vec{c}|^2 = |\vec{a} - \vec{b}|^2 + |\vec{c} - \vec{b}|^2 - 2|\vec{a} - \vec{b}||\vec{c} - \vec{b}|\cos(\angle L(\vec{a})L(\vec{b})L(\vec{c})).$$

Now compare this with the law of cosines for $\triangle a\vec{b}\vec{c}$:

$$|\vec{a} - \vec{c}|^2 = |\vec{a} - \vec{b}|^2 + |\vec{c} - \vec{b}|^2 - 2|\vec{a} - \vec{b}||\vec{c} - \vec{b}|\cos(\angle a\vec{b}\vec{c}).$$

From these equations, we conclude that:

$$\cos(\angle L(\vec{a})L(\vec{b})L(\vec{c})) = \cos(\angle a\vec{b}\vec{c}).$$

Since the cosine function is one-to-one in the range $[0°, 180°]$, it follows that:

$$\angle L(\vec{a})L(\vec{b})L(\vec{c}) = \angle \vec{a}\vec{b}\vec{c}.$$

Similarly, we can show that:

$$\angle L(\vec{b})L(\vec{c})L(\vec{a}) = \angle \vec{b}\vec{c}\vec{a}, \quad \angle L(\vec{c})L(\vec{a})L(\vec{b}) = \angle \vec{c}\vec{a}\vec{b}.$$

$$\square$$

In studying symmetries, and in particular, isometries, it is often helpful to break down a transformation into simpler components, viewing it as a composition of fundamental transformations. This approach not only simplifies the analysis but also reveals the underlying structure and properties of the isometry.

The simplest examples of similarities are the *dilations* centered at the origin. A dilation with magnification factor $m \neq 0$ is represented by the matrix

$$\mathbf{M}_m = \begin{bmatrix} m & 0 \\ 0 & m \end{bmatrix}, \qquad \mathbf{M}_m(\vec{v}) = m\vec{v}.$$

To verify that these are in fact similarities, consider two points $\vec{p}$ and $\vec{q}$:

$$\left| \mathbf{M}_m(\vec{p}) - \mathbf{M}_m(\vec{q}) \right| = \left| m\vec{p} - m\vec{q} \right| = m\,|\vec{p} - \vec{q}|.$$

Thus, all lengths are multiplied by the same factor m, and the transformation is a similarity.

As we progress through this section, we will examine more complex similarities and investigate how combinations of various similarities and isometries can be used to construct increasingly intricate transformations.

To compose two linear transformations T_1 and T_2, we apply one after the other. If T_1 maps a vector $\vec{v}$ to $T_1(\vec{v})$, and T_2 maps a vector $\vec{w}$ to $T_2(\vec{w})$, then their composition is given by

$$(T_2 \circ T_1)(\vec{v}) = T_2\big(T_1(\vec{v})\big).$$

It is important to note that the order of multiplication matters, since matrix multiplication is not commutative: in most cases, $\mathbf{BA} \neq \mathbf{AB}$.

The following theorem summarizes some fundamental properties of isometries and similarities. These results will be central as we explore these transformations in greater depth.

Theorem 8.3

(1) Translations are isometries.

(2) The composition of a similarity L with magnification m and a similarity K with magnification n is a similarity $K \circ L$ with magnification mn

(3) The linear transformation $L(\vec{v}) = \mathbf{A}\vec{v} + w$ is a similarity with magnification m if and only if the matrix linear transformation $K(\vec{v}) = \mathbf{A}\vec{v}$ is a similarity with magnification m.

(4) If the matrix linear transformation $L(\vec{v}) = \mathbf{A}\vec{v}$ is a similarity with magnification m, then $L = M \circ K$ where $M(\vec{v}) = \mathbf{M}\vec{v}$, $\mathbf{M} = \begin{bmatrix} m & 0 \\ 0 & m \end{bmatrix}$ and K is an isometry.

Proof

(1) Consider the translation $T(\vec{v}) = \vec{v} + \vec{w}$, and let $\vec{p}$ and $\vec{q}$ be any two points. We must show that the distance between $T(\vec{p})$ and $T(\vec{q})$ is the same as the distance between $\vec{p}$ and $\vec{q}$. We have:

$$|T(\vec{p}) - T(\vec{q})| = |(\vec{p} + \vec{w}) - (\vec{q} + \vec{w})| = |\vec{p} - \vec{q}|.$$

(2) Let $\vec{p}$ and $\vec{q}$ be any two points. Then

$$|L(\vec{p}) - L(\vec{q})| = m|\vec{p} - \vec{q}| = |m\vec{p} - m\vec{q}|.$$

But then

$$|(K \circ L)(\vec{q}) - (K \circ L)(\vec{q})| = n|L(\vec{p}) - L(\vec{q})| = nm|\vec{p} - \vec{q}|.$$

(3) Assume that $K(\vec{v}) = \mathbf{A}\vec{v}$ a similarity with magnification m and T is the translation $T(\vec{v}) = \vec{v} + \vec{w}$. By 1), T a similarity with magnification 1. Then $L(\vec{v}) = \mathbf{A}\vec{v} + \vec{w}$ is the composition $L = T \circ K$ and, by (2), it is a similarity with magnification $m \times 1 = m$.

Conversely, assume that $L(\vec{v}) = \mathbf{A}\vec{v} + \vec{w}$ is a similarity with magnification m and compose it with the translation $T'(\vec{v}) = \vec{v} - \vec{w}$ to see that $K = T' \circ L$ is a similarity of the form $K(\vec{v}) = \mathbf{A}\vec{v}$ with magnification m and that $L = T \circ K$.

(4) One easily checks that $\mathbf{M}^{-1} = \begin{bmatrix} \frac{1}{m} & 0 \\ 0 & \frac{1}{m} \end{bmatrix}$ and that the matrix linear transformation $\mathbf{M}^{-1}(\vec{v}) = \mathbf{M}^{-1}\vec{v}$ is a dilation with magnification $\frac{1}{m}$. Then K, the composition $K(\vec{v}) = (\mathbf{M}^{-1} \circ L)(\vec{v}) = (\mathbf{M}^{-1}\mathbf{A})(\vec{v})$ is an isometry. Furthermore, $\mathbf{M} \circ K = \mathbf{M} \circ (\mathbf{M}^{-1} \circ L) = L$.

$\square$

In view of Part (4) of the theorem, will concentrate on the isometries. Of these, the translations are the simplest. The effect of translations on the plane is straightforward: points are shifted the same distance in the same direction. This is equivalent to moving the object by a fixed amount along the x-axis and then by another fixed amount along the y-axis. These isometries can be expressed as $T(\vec{v}) = \vec{v} + \vec{w}$, where $\vec{w}$ represents the translation vector.

We start with a simple, real-world example to illustrate how translations work. Imagine Sarah Jane is designing a garden and has already built a raised bed. After careful planning, she realizes the entire structure needs to shift two feet to the left and two feet down to avoid a large tree. To ensure the garden bed remains parallel to the sides of her fence, she must move it without any rotation. This process is analogous to a two-dimensional translation: the shape, size, and orientation of the garden bed stay the same, but its position changes.

The translation vector for this situation is given by $\vec{w} = \begin{bmatrix} -2 \\ -2 \end{bmatrix}$ indicating a shift two units left and two units down. In the figure below, the original garden bed is shown with solid lines, while the new position is outlined with dashed lines. Each point in the solid figure is translated exactly two units left and two units down to create the dashed image.

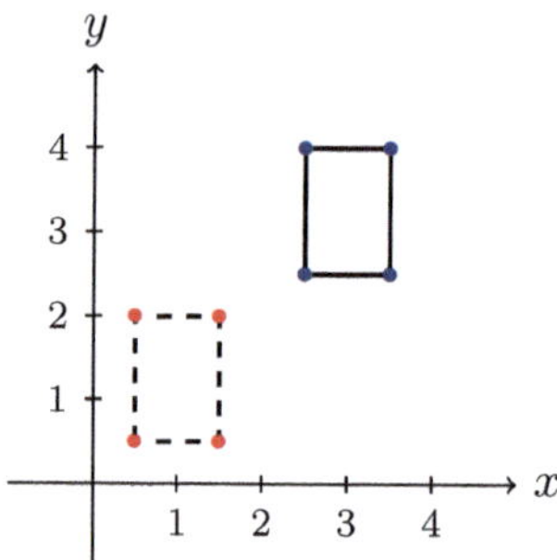

Exercise 8.9 Graph the translation of the triangle with vertices at $\vec{a} = \begin{bmatrix} -1 \\ 2 \end{bmatrix}$, $\vec{b} = \begin{bmatrix} 0 \\ 2 \end{bmatrix}$ and $\vec{c} = \begin{bmatrix} 0 \\ 3 \end{bmatrix}$ by the vector $\vec{v} = \begin{bmatrix} 3 \\ -1 \end{bmatrix}$.

Exercise 8.10 Find the translation matrix representing a shift of left by six units and up five units. Compute the new position of $\vec{r} = \begin{bmatrix} 3 \\ -4 \end{bmatrix}$ after applying this translation.

Exercise 8.11 Consider the linear transformation $T(\vec{v}) = \vec{v} + \vec{t}$ where

$$\vec{t} = \begin{bmatrix} 2 \\ -3 \end{bmatrix}.$$

a. Determine if there are any fixed points under this transformation.

b. Discuss the conditions under which a linear translation can have fixed points.

Now, let's explore a more involved example that still relies on translations. Consider the following problem involving two neighboring small towns, A and B, connected by a direct road (depicted in red below). Recently, due to an increase in heavy rains, a city just west of these towns has proposed building a wide (100 feet) but shallow drainage ditch extending from the city to the river east of the towns. This drainage ditch will pass between the two towns. As a result, the existing road will need to be replaced with two segments connected by a bridge. A potential route for this new road is shown in green.

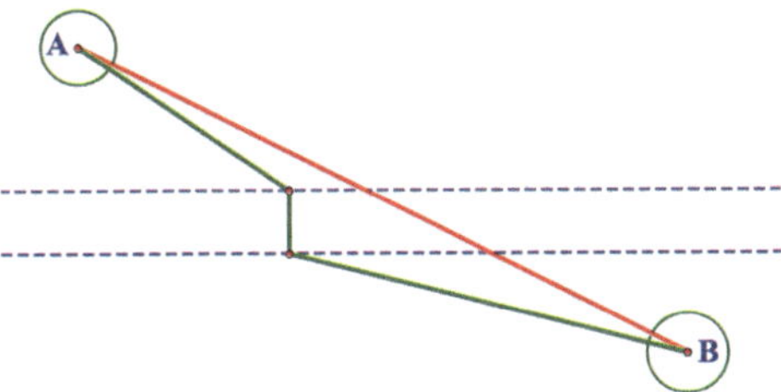

Since the old road is in poor condition, the road will need to be rebuilt and this reconstruction is an opportunity to optimize the new route. The question is: Where should the bridge be placed to minimize the total road length? To determine the optimal bridge location, we use a translation. Specifically, we "translate" town B north by the width of the ditch and identify the direct route from town A to the translated point B'.

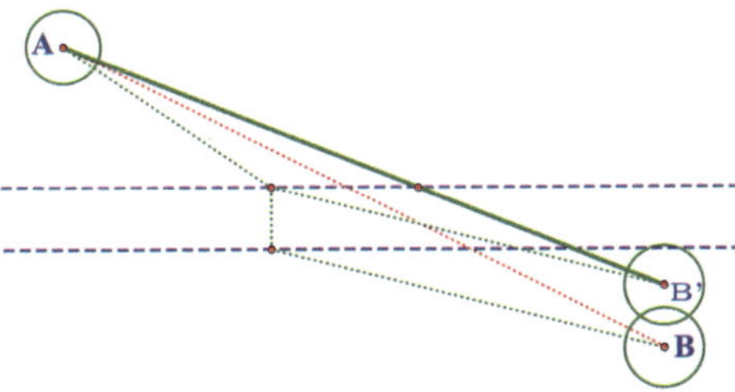

The north entrance to the bridge is located at the intersection of this straight route with the northern edge of the ditch. This point determines where the bridge should cross to ensure the total road length is minimized.

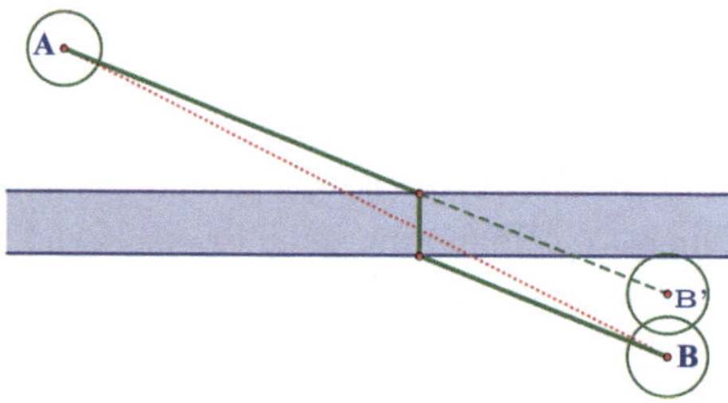

Exercise 8.12 Create your own exercises: one where two parallel ditches separate points A and B, and another where two non-parallel ditches separate A and B. Then, solve these "bridge placement" problems.

Two-Dimensional Isometries

9

9.1 Matrix Isometries

To further understand isometries in two-dimensions, consider an isometry in its general form

$$L(\vec{v}) = \mathbf{A}\vec{v} + \vec{w}.$$

Translations arise as a special case when $\mathbf{A}$ is the identity matrix.

If we temporarily set aside the translation component, $\vec{w}$, the transformation takes the simpler form

$$L(\vec{v}) = \mathbf{A}\vec{v}.$$

We refer to isometries of this form as *matrix isometries.*

In two-dimensions, matrix isometries consist of *rotations* and *reflections*, along with their compositions. By focusing on these fundamental linear components, we uncover the rotational and reflectional structure that underlies any isometry.

Developing a solid understanding of matrix isometries before reintroducing translations allows us to analyze more effectively how these linear transformations interact with translations.

Let $L(\vec{v}) = \mathbf{A}v$, where

$$\mathbf{A} = \begin{bmatrix} a_{11} & a_{12} \\ a_{21} & a_{22} \end{bmatrix}$$

be a matrix isometry. The first column of $\mathbf{A}$ represents the image of the unit vector $\vec{x}$, and the second column of $\mathbf{A}$ represents the image of the unit vector $\vec{y}$. Specifically:

$$\mathbf{A}\vec{x} = \begin{bmatrix} a_{11} & a_{12} \\ a_{21} & a_{22} \end{bmatrix} \begin{bmatrix} 1 \\ 0 \end{bmatrix} = \begin{bmatrix} a_{11} \\ a_{21} \end{bmatrix} \quad \text{and} \quad \mathbf{A}\vec{y} = \begin{bmatrix} a_{11} & a_{12} \\ a_{21} & a_{22} \end{bmatrix} \begin{bmatrix} 0 \\ 1 \end{bmatrix} = \begin{bmatrix} a_{12} \\ a_{22} \end{bmatrix}.$$

We begin with a theorem that allows us to identify the possible forms of matrix $\mathbf{A}$ that will ensure L is an isometry.

Theorem 9.1 *If the linear transformation* $L(\vec{v}) = \mathbf{A}\vec{v}$ *is an isometry then either* $\mathbf{A} = \begin{bmatrix} \cos\theta & -\sin\theta \\ \sin\theta & \cos\theta \end{bmatrix}$ *or* $\mathbf{A} = \begin{bmatrix} \cos\theta & \sin\theta \\ \sin\theta & -\cos\theta \end{bmatrix}$, *for some angle* θ *in the range* $0 \leq \theta \leq 2\pi$.

Proof Let $\mathbf{A} = \begin{bmatrix} a_{11} & a_{12} \\ a_{21} & a_{22} \end{bmatrix}$. The first column vector of the matrix $\mathbf{A}$ is the image of unit vector on the x-axis and, therefore since L is an isometry $\begin{bmatrix} a_{11} \\ a_{21} \end{bmatrix}$ must have length one. So we must have that $\begin{bmatrix} a_{11} \\ a_{21} \end{bmatrix}$ is a segment from the origin that terminates on the unit circle and therefore it's coordinates are $\begin{bmatrix} \cos(\theta) \\ \sin(\theta) \end{bmatrix}$, where θ is the counterclockwise angle it makes with the positive x-axis.

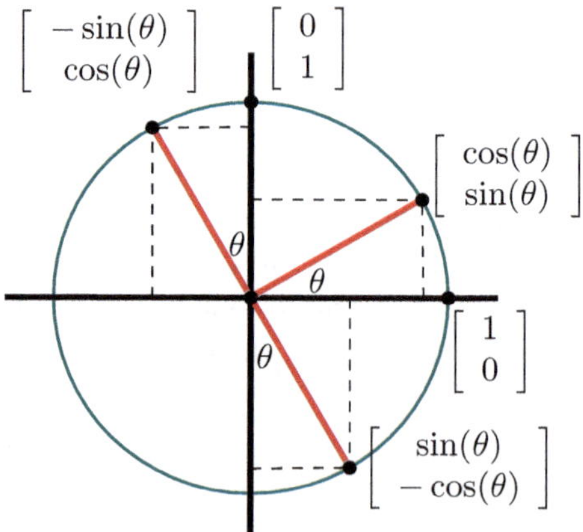

Since L is an isometry, angles are preserved. Therefore, since $\vec{x}$ and $\vec{y}$ are orthogonal, their images, which are the first and second column vectors of $\mathbf{A}$, must also be orthogonal.

Assuming that $\begin{bmatrix} a_{12} \\ a_{22} \end{bmatrix}$ is orthogonal to $\begin{bmatrix} \cos(\theta) \\ \sin(\theta) \end{bmatrix}$ and has unit length, there are only two possible choices for the second column: $\begin{bmatrix} -\sin(\theta) \\ \cos(\theta) \end{bmatrix}$ or $\begin{bmatrix} \sin(\theta) \\ -\cos(\theta) \end{bmatrix}$. We have illustrated this proof for θ in the first quadrant, but the proof is valid for all θ. $\qquad\square$

We will adopt the notation $\mathbf{D}_\theta = \begin{bmatrix} \cos\theta & -\sin\theta \\ \sin\theta & \cos\theta \end{bmatrix}$ and $\mathbf{O}_\theta = \begin{bmatrix} \cos\theta & \sin\theta \\ \sin\theta & -\cos\theta \end{bmatrix}$. In the example pictured above, θ equals 30 degrees and we see that $\mathbf{D}_{30°}$ rotates the unit x vector, $\begin{bmatrix} 1 \\ 0 \end{bmatrix}$, by 30° onto the vector $\begin{bmatrix} \cos\theta \\ \sin\theta \end{bmatrix}$ and rotates the unit y vector, $\begin{bmatrix} 0 \\ 1 \end{bmatrix}$, by 30° onto the vector $\begin{bmatrix} -\sin\theta \\ \cos\theta \end{bmatrix}$ while $\mathbf{O}_{30°}$ reflects the unit x vector through the 15° bisector of the angle $\left(\begin{bmatrix} 0 \\ 1 \end{bmatrix}, \begin{bmatrix} 0 \\ 0 \end{bmatrix}, \begin{bmatrix} \cos\theta \\ \sin\theta \end{bmatrix} \right)$ onto the vector $\begin{bmatrix} \cos\theta \\ \sin\theta \end{bmatrix}$ and reflects the unit y vector $\begin{bmatrix} 0 \\ 1 \end{bmatrix}$ through this same bisector onto the vector $\begin{bmatrix} -\sin\theta \\ \cos\theta \end{bmatrix}$.

Exercise 9.1 Let $\triangle$ be the triangle with vertices Let $\vec{a} = \begin{bmatrix} 2 \\ 3 \end{bmatrix}$ and $\vec{b} = \begin{bmatrix} 5 \\ 7 \end{bmatrix}$ and $\vec{c} = \begin{bmatrix} 5 \\ 3 \end{bmatrix}$. Let $\theta = \frac{\pi}{6}$.

(1) Compute the image of $\triangle$ under the isometry $\mathbf{D}_\theta$.
(2) Compute the image of $\triangle$ under the isometry $\mathbf{O}_\theta$.
(3) Identify any feature(s) you observe that distinguish the actions of $\mathbf{D}_\theta$ and $\mathbf{O}_\theta$ on this triangle.

In general, the determinant of the matrix associated with a two-dimensional linear function provides valuable insights into the transformation's behavior. Similar to the slope in a one-dimensional linear function, the determinant reveals key properties. If a is the slope of the linear function f, then $|a|$ is the factor by which all distances are scaled, and the sign of a indicates whether f preserves or reverses orientation. Analogously, for a 2D linear function $L(\vec{v}) = \mathbf{A}\vec{v}$, the value $\det(\mathbf{A})$ represents the factor by which all areas are scaled, while the sign of $\det(\mathbf{A})$ tells us if L preserves or reverses orientation.

Observe that $\det(\mathbf{D}_\theta) = 1$ and $\det(\mathbf{O}_\theta) = -1$. Thus, isometries with a $\mathbf{D}_\theta$ matrix will preserve orientation, and we refer to these as *direct isometries*, whereas isometries with a $\mathbf{O}_\theta$ matrix will reverse orientation, and we call these *opposite isometries*. Here, $\mathbf{D}_\theta$ represents a rotation matrix that rotates vectors counterclockwise by an angle θ around the origin, while $\mathbf{O}_\theta$ is a reflection matrix.

9.2   Isometry Classifications

Isometries in the Euclidean plane fall into two fundamental categories: direct and opposite isometries. Both can be visualized as ways of moving a shape around a flat surface without altering its size or form.

Direct isometries preserve orientation, they are like sliding a sheet of paper across a table without lifting or flipping it. These transformations include translations and rotations, both of which maintain side lengths, angles, and orientation of the figure. We will show that any movement in the Euclidean plane that maintains these properties involves only rotations and translations.

Opposite isometries, on the other hand, reverse orientation. They are akin to flipping the sheet of paper over before moving it. This category includes reflections and glide reflections, where a glide reflection combines a reflection across a line with a translation along that line. We should point out that, since an opposite isometry reverses orientation, the composition of two opposite isometries reverses orientation twice and therefore is a direct isometry.

Exercise 9.2 Figure 9.1 shows five triangles labeled Y, B, R, P, and G. Each of these triangles is congruent to the others. Determine which pairs of triangles are related by direct isometries and which pairs are related by opposite isometries.

It is intuitive to see that translations have no fixed points, rotations have a single fixed point, reflections have a single point-wise fixed line and glide reflections have single fixed (but not point-wise) line. The following theorem will unify these observations into a single result. However, before we can prove that theorem, we need to two lemmas that will be used in the proof.

Lemma 9.1 *Let L be an isometry.*

(1) If L has two distinct fixed points, then either L is the identity or L is the reflection through the line containing the two fixed points.

(2) If L^2 is the identity, then L is one of the following:

Fig. 9.1 Triangles
transformed by isometries

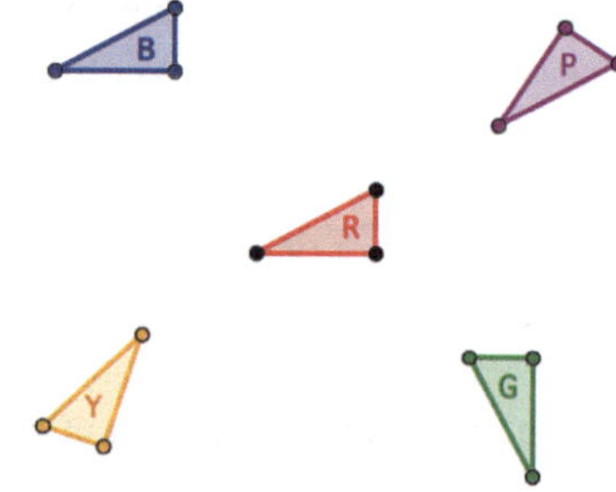

(a) the identity

(b) a 180 ° rotation about some point

(c) a reflection through some line

Proof

(1) Let $\vec{p}$ and $\vec{q}$ be two distinct fixed points for L, and let ℓ be the unique line through $\vec{p}$ and $\vec{q}$. Since the restriction of L to ℓ is a linear function on ℓ with two distinct fixed points, it follows that the restriction of L to ℓ is the identity on ℓ. Thus, ℓ is a pointwise fixed line for L.

Now, let $\vec{r}$ be any point not on ℓ. If $\vec{r}$ is also fixed by L, then L agrees with the identity on three non-collinear points: $\vec{p}$, $\vec{q}$, and $\vec{r}$. Therefore, L must be the identity transformation.

If $L(\vec{r}) \neq \vec{r}$, then $L(\vec{r})$ must be the reflection of $\vec{r}$ through ℓ. To see this, note that the distance from $L(\vec{r})$ to the fixed point $\vec{p}$ must equal the distance from $\vec{r}$ to $\vec{p}$. That is, $L(\vec{r})$ lies on the circle centered at $\vec{p}$ and passing through $\vec{r}$. Similarly, $L(\vec{r})$ lies on the circle centered at $\vec{q}$ and passing through $\vec{r}$.

By basic geometry, ℓ is the perpendicular bisector of the segment joining $\vec{r}$ and $L(\vec{r})$, and the reflection through ℓ interchanges $\vec{r}$ and $L(\vec{r})$. Finally, this reflection agrees with L on three non-collinear points: $\vec{p}$, $\vec{q}$, and $\vec{r}$. Therefore, it follows that L is this reflection.

(2) Now assume that L^2 is the identity. Clearly, L could be the identity. If L is not the identity, let $\vec{p}$ be a point that is not fixed, and let $\vec{q} = L(\vec{p})$.

Since $L(\vec{q}) = L^2(\vec{p}) = \vec{p}$, it follows that L interchanges $\vec{p}$ and $\vec{q}$, and maps the segment $\overline{pq}$ onto itself. Hence, $\vec{c}$, the midpoint of the $\overline{pq}$ segment, must be a fixed point of L.

Now, let $\vec{r}$ be a point, other than $\vec{c}$, on ℓ, the perpendicular bisector of the segment $\overline{pq}$. Since L is an isometry, the distances from $L(\vec{r})$ to $\vec{p}$, $\vec{c}$, and $\vec{q}$ must be the same as the distances from $\vec{r}$ to $\vec{p}$, $\vec{c}$, and $\vec{q}$. Therefore, the only possibilities for $L(\vec{r})$ are $L(\vec{r}) = \vec{r}$, or $L(\vec{r})$ is the reflection of $\vec{r}$ through the segment $\overline{pq}$, which is equivalent to the 180° rotation of $\vec{r}$ about $\vec{c}$.

In the first case, L agrees with the reflection through ℓ on three non-collinear points: $\vec{p}$, $\vec{q}$, and $\vec{r}$. In the second case, L agrees with the 180° rotation about $\vec{c}$ (which interchanges $\vec{p}$ and $\vec{q}$) on the same three non-collinear points: $\vec{p}$, $\vec{q}$, and $\vec{r}$. $\quad\square$

In the next lemma, we consider the simplest class of opposite isometries are the matrix isometries of the form $L(\vec{v}) = \mathbf{O}_\theta \vec{v}$.

Lemma 9.2 $L(\vec{v}) = O_\theta \vec{v}$ *is the reflection through the line ℓ through the origin making a $\frac{1}{2}\theta°$ counterclockwise angle from the positive x-axis and is given parmetrically by* $\lambda \begin{bmatrix} 1 + \cos\theta \\ \sin\theta \end{bmatrix}$.

Proof We have by direct computation $L(\vec{0}) = \vec{0}$, $L(\begin{bmatrix} 1 \\ 0 \end{bmatrix}) = \begin{bmatrix} \cos\theta \\ \sin\theta \end{bmatrix}$, and $L(\begin{bmatrix} \cos\theta \\ \sin\theta \end{bmatrix}) = \begin{bmatrix} 1 \\ 0 \end{bmatrix})$. Since $\begin{bmatrix} 1 \\ 0 \end{bmatrix}$ and $\begin{bmatrix} \cos\theta \\ \sin\theta \end{bmatrix}$ are interchanged, the $\vec{m}$, midpoint of the segment joining them, is on the axis of reflection ℓ. Hence, ℓ is the line through the origin and $\vec{m}$, which is easily seen to be the bisector of the angle $\angle(\begin{bmatrix} 1 \\ 0 \end{bmatrix}, \vec{0}, \begin{bmatrix} \cos\theta \\ \sin\theta \end{bmatrix})$. Finally since the vector $\begin{bmatrix} \cos\theta \\ \sin\theta \end{bmatrix}$ makes a counterclockwise angle of θ from the positive x axis, ℓ makes a counterclockwise angle of $\frac{1}{2}\theta$ from the positive x axis. Hence ℓ is given parametrically by $\lambda \begin{bmatrix} \cos(\frac{1}{2}\theta) \\ \sin(\frac{1}{2}\theta) \end{bmatrix}$. Using the half-angle trig formulas, we could see that this simplifes to $\lambda \begin{bmatrix} 1 + \cos\theta \\ \sin\theta \end{bmatrix}$. Alternatively, we can simply check that the vector is fixed by L.

$$L(\lambda \begin{bmatrix} 1 + \cos\theta \\ \sin\theta \end{bmatrix}) = \lambda \begin{bmatrix} \cos\theta & \sin\theta \\ \sin\theta & -\cos\theta \end{bmatrix} \begin{bmatrix} 1 + \cos\theta \\ \sin\theta \end{bmatrix}.$$ Expanding the matrix product, we get:

$$L(\lambda \begin{bmatrix} 1 + \cos\theta \\ \sin\theta \end{bmatrix}) = \lambda \begin{bmatrix} \cos\theta(1 + \cos\theta) + \sin\theta\sin\theta \\ \sin\theta(1 + \cos\theta) - \cos\theta\sin\theta \end{bmatrix}$$

$$= \lambda \begin{bmatrix} \cos\theta + \cos^2\theta + \sin^2\theta \\ \sin\theta + \cos\theta\sin\theta - \cos\theta\sin\theta \end{bmatrix}.$$

This simplifies to: $L\left(\lambda \begin{bmatrix} 1 + \cos\theta \\ \sin\theta \end{bmatrix}\right) = \lambda \begin{bmatrix} 1 + \cos\theta \\ \sin\theta \end{bmatrix}$. Thus, every point on the line ℓ is fixed by L. Therefore, ℓ is the axis of reflection for L. $\square$

Exercise 9.3 Find O_θ so that $L(\vec{v}) = O_\theta \vec{v}$ reflects points across the line $y = x$.

Exercise 9.4 Consider the triangle $\Delta(\vec{p}, \vec{q}, \vec{r})$ where

$$\vec{p} = \begin{bmatrix} 1 \\ 2 \end{bmatrix}, \vec{q} = \begin{bmatrix} 4 \\ 3 \end{bmatrix}, \text{ and } \vec{r} = \begin{bmatrix} 2 \\ 5 \end{bmatrix}.$$

a. Reflect this triangle across the line $\vec{y} = -\vec{x}$ and find the new coordinates.

b. Plot the original and reflected triangles on the coordinate plane.

c. Show that the distance between any two points in the original figure is the same as the distance between their reflected images.

We would like to develop systematic methods for deriving formulas for specific linear transformations. For instance, how can we determine the formula for a $90°$ rotation about the point $\begin{bmatrix} 2 \\ 2 \end{bmatrix}$, or the formula for a reflection across the line $y = 2x + 1$? The following theorem provides a key step toward achieving this goal. Building on the results from Lemmas 9.1 and 9.2, we now have the necessary tools to prove the following isometry classification theorem.

Theorem 9.2 *Let $L(\vec{v}) = \mathbf{A}\vec{v} + \vec{w}$ be an isometry.*

(1) If L is a direct isometry, then L is either a translation or rotation. If L is a translation, then it has no fixed points and L has the form $L(\vec{v}) = \vec{v} + \vec{w}$. If L is a rotation, it is a counterclockwise rotation by the fixed angle $\theta°$ about the fixed point c. In this case, $\vec{c} = (\mathbf{I} - \mathbf{D}_\theta)^{-1}\vec{w}$ and L has the form $L(\vec{v}) = \mathbf{D}_\theta + (\mathbf{I} - \mathbf{D}_\theta)\vec{c}$.

(2) If L is an opposite isometry, it does not have a unique fixed point; instead, it will either have a line of fixed points (the axis of a reflection) or a fixed line that is not point-wise fixed (the axis of a glide reflection). Since L is opposite, $L(\vec{v}) = \mathbf{O}_\theta(v) + \vec{w}$ for some angle θ and some vector $\vec{w}$.

a. If $\vec{w} = \vec{0}$, L is the reflection through the line ℓ passing through $\vec{0}$, and making a counterclockwise angle of $\frac{1}{2}\theta$ degrees with the positive x-axis and is given parametrically: $\ell = \lambda \begin{bmatrix} 1 + \cos\theta \\ \sin\theta \end{bmatrix}$.

b. If $\vec{w} \neq \vec{0}$ and $L(\vec{w}) = 0$, then L is the reflection through the perpendicular bisector of the segment joining 0 and $\vec{w}$.

c. If $\vec{w} \neq 0$ and $L(\vec{w}) \neq 0$, then L is the glide reflection with the line containing the points $\frac{1}{2}\vec{w}$ and $L(\frac{1}{2}\vec{w})$ as the axis of the reflection to be followed by the translation $T(\vec{v}) = \vec{v} + \frac{1}{2}(\mathbf{O}_\theta\vec{w} + \vec{w})$.

Proof

(1) Suppose that L is a direct isometry, with $L(\vec{v}) = \mathbf{D}_\theta\vec{v} + \vec{w}$. By Theorem 6.3, L will have a unique fixed point, $\vec{c} = (\mathbf{I} - \mathbf{D}_\theta)^{-1}\vec{w}$, if and only if the matrix $\mathbf{I} - \mathbf{D}_\theta$ has

an inverse. This occurs if and only if its determinant is non-zero. We compute the determinant as follows:

$$\det(\mathbf{I} - \mathbf{D}_\theta) = \det\begin{bmatrix} 1 - \cos\theta & -\sin\theta \\ \sin\theta & 1 - \cos\theta \end{bmatrix} = (1 - \cos\theta)^2 + (\sin\theta)^2 = 2(1 - \cos\theta).$$

Clearly, $\det(\mathbf{I} - \mathbf{D}_\theta) = 0$ if and only if $\cos\theta = 1$, which happens only when $\theta = 0$ or $\theta = 2\pi$. In either case, $\mathbf{D}_\theta = \mathbf{I}$, and L is a translation. Otherwise, L has a fixed point and is the rotation: $L(\vec{v}) = \mathbf{D}_\theta\vec{v} + (\mathbf{I} - \mathbf{D}_\theta)\vec{c}$.

(2) Now let L be an opposite isometry of the form

$$L(\vec{v}) = \mathbf{O}_\theta\vec{v} + \vec{w}.$$

The map L has a unique fixed point

$$\vec{c} = (\mathbf{I} - \mathbf{O}_\theta)^{-1}\vec{w}$$

if and only if the matrix $\mathbf{I} - \mathbf{O}_\theta$ is invertible. However, computing its determinant shows that this never occurs since the determinant is 0 for all values of θ:

$$\det(\mathbf{I} - \mathbf{O}_\theta) = \det\begin{bmatrix} 1 - \cos\theta & -\sin\theta \\ -\sin\theta & 1 + \cos\theta \end{bmatrix} = 1 - \cos^2\theta - \sin^2\theta = 0.$$

Thus, an opposite isometry cannot have a unique fixed point. This does not rule out the possibility of having a point-wise fixed line.

Before considering the individual cases, note that

$$\mathbf{O}_\theta^2 = \mathbf{I} \quad \text{for all } \theta.$$

It follows that L^2 is a translation. More precisely,

$$L^2(\vec{v}) = L(L(\vec{v})) = L(\mathbf{O}_\theta\vec{v} + \vec{w}) = \mathbf{O}_\theta^2\vec{v} + \mathbf{O}_\theta\vec{w} + \vec{w} = \vec{v} + (\mathbf{O}_\theta\vec{w} + \vec{w})$$

shows that L^2 is a translation by the vector $L(\vec{w}) = \mathbf{O}_\theta\vec{w} + \vec{w}$

(a) $\vec{w} = \vec{0}$. Then $L(\vec{v}) = \mathbf{O}_\theta\vec{v}$ and the origin is fixed since $L(\vec{0}) = \vec{0}$. Furthermore, L^2 is the identity map. Now we have

$$L\begin{bmatrix} 1 \\ 0 \end{bmatrix} = \begin{bmatrix} \cos\theta \\ \sin\theta \end{bmatrix},$$

and since L^2 is the identity,

$$L \begin{bmatrix} \cos\theta \\ \sin\theta \end{bmatrix} = L\left(L \begin{bmatrix} 1 \\ 0 \end{bmatrix} \right) = L^2 \begin{bmatrix} 1 \\ 0 \end{bmatrix} = \begin{bmatrix} 1 \\ 0 \end{bmatrix}.$$

Thus, L interchanges the endpoints of the segment joining $\begin{bmatrix} 1 \\ 0 \end{bmatrix}$ and $\begin{bmatrix} \cos\theta \\ \sin\theta \end{bmatrix}$, fixing the midpoint $\vec{m}$. Since L also fixes $\vec{0}$, the line ℓ through $\vec{0}$ and $\vec{m}$ is fixed point-wise, making it the axis of reflection. Since $\vec{m}$ is the bisector of the segment $\begin{bmatrix} 1 \\ 0 \end{bmatrix} - \begin{bmatrix} \cos\theta \\ \sin\theta \end{bmatrix}$, ℓ is the bisector of the angle $\begin{bmatrix} 1 \\ 0 \end{bmatrix}, \begin{bmatrix} 0 \\ 0 \end{bmatrix}, \begin{bmatrix} \cos\theta \\ \sin\theta \end{bmatrix}$ and therefore makes the counterclockwise angle of $\frac{\theta}{2}$ with the x-axis. The parametric equation for this axis is $\ell = \lambda \begin{bmatrix} \cos\left(\frac{\theta}{2}\right) \\ \sin\left(\frac{\theta}{2}\right) \end{bmatrix}$. Using the half-angle trig formulas, we can simplify this to $\lambda \begin{bmatrix} 1 + \cos\theta \\ \sin\theta \end{bmatrix}$: however, by Lemma 9.2 we can skip that computation.

(b) Next, assume $\vec{w} \neq \vec{0}$. Consider the point $\frac{1}{2}\vec{w}$, the midpoint of the segment joining $\vec{0}$ and $L(\vec{0}) = \vec{w}$. The midpoint of the segment joining $\vec{w}$ and $L(\vec{w})$ is

$$\frac{1}{2}(\vec{w} + L(\vec{w})) = \frac{1}{2}(\vec{w} + \mathbf{O}_\theta \vec{w} + \vec{w}) = \frac{1}{2}\mathbf{O}_\theta \vec{w} + \vec{w} = \mathbf{O}_\theta\left(\tfrac{1}{2}\vec{w}\right) + \vec{w} = L\left(\tfrac{1}{2}\vec{w}\right).$$

We now consider two subcases.

Case 1: $L(\vec{w}) = \vec{0}$. In this case, $\frac{1}{2}\vec{w}$ is a fixed point of L: since L interchanges $\vec{0}$ and $\vec{w}$, it maps the $\vec{0}$-$\vec{w}$ segment onto itself thereby mapping the midpoint onto itself; which we can verify directly:

$$L\left(\tfrac{1}{2}\vec{w}\right) = \mathbf{O}_\theta\left(\tfrac{1}{2}\vec{w}\right) + \vec{w} = \tfrac{1}{2}(\mathbf{O}_\theta \vec{w} + \vec{w}) + \tfrac{1}{2}\vec{w} = L(\vec{w}) + \tfrac{1}{2}\vec{w} = \vec{0} + \tfrac{1}{2}\vec{w}.$$

Since an opposite isometry cannot have a unique fixed point, there must be at least one other fixed point, say $\vec{u}$. Let ℓ be the line through $\frac{1}{2}\vec{w}$ and $\vec{u}$. The restriction of L to ℓ is a linear transformation of ℓ with two fixed points, $\frac{1}{2}\vec{w}$ and $\vec{u}$, so it must be the identity map on ℓ. In addition, ℓ is the perpendicular bisector of the segment $\vec{0} - \vec{w}$. To see this, note that the angle $\angle(\vec{0}, \frac{1}{2}\vec{w}, \vec{u})$ is mapped by L onto $\angle(\vec{w}, \frac{1}{2}\vec{w}, \vec{u})$, and these two angles must be equal. Since $\angle(\vec{0}, \frac{1}{2}\vec{w}, \vec{w})$ is a straight angle of $180°$, it follows that $\angle(\vec{0}, \frac{1}{2}\vec{w}, \vec{u})$ and $\angle(\vec{w}, \frac{1}{2}\vec{w}, \vec{u})$ are each $90°$. Thus, L agrees with the reflection through ℓ at three non-collinear points: $\vec{0}$, $\frac{1}{2}\vec{w}$, and $\vec{u}$. Therefore, L is the reflection through ℓ.

Case 2: $L(\vec{w}) \neq \vec{0}$. Then $\frac{1}{2}\vec{w}$ is not a fixed point. Instead, L translates $\frac{1}{2}\vec{w}$ by the vector $\frac{1}{2}(\mathbf{O}_\theta \vec{w} + \vec{w})$:

$$L(\tfrac{1}{2}\vec{w}) = \mathbf{O}_\theta\left(\tfrac{1}{2}\vec{w}\right) + \vec{w} = (\tfrac{1}{2}\mathbf{O}_\theta\vec{w} + \vec{w}) = \tfrac{1}{2}(\mathbf{O}_\theta\vec{w} + \vec{w}) + \tfrac{1}{2}\vec{w}$$

Applying L again translates $L(\tfrac{1}{2}\vec{w})$ onto $L^2(\tfrac{1}{2}\vec{w})$ by the same vector

$$\tfrac{1}{2}(\mathbf{O}_\theta\vec{w} + \vec{w}) + L\left(\tfrac{1}{2}\vec{w}\right) = \tfrac{1}{2}(\mathbf{O}_\theta\vec{w} + \vec{w}) + \tfrac{1}{2}(\mathbf{O}_\theta\vec{w} + \vec{w}) + \tfrac{1}{2}\vec{w}$$

$$= (\mathbf{O}_\theta\vec{w} + \vec{w}) + \tfrac{1}{2}\vec{w}$$

$$= L^2\left(\tfrac{1}{2}\vec{w}\right).$$

We conclude that L restricted to ℓ is the translation $\tfrac{1}{2}(\mathbf{O}_\theta\vec{w} + \vec{w})$.

Finally we note that composing L with the translation $-\tfrac{1}{2}(\mathbf{O}_\theta\vec{w} + \vec{w})$ fixes the line ℓ and is, hence the reflection through the line ℓ. Therefore L is the glide reflection consisting of the reflection across ℓ followed by the translation by the vector $\tfrac{1}{2}(\mathbf{O}_\theta\vec{w} + \vec{w})$. $\qquad\square$

To illustrate this last part of the proof, we use a specific glide reflection.

Let $L(\vec{v}) = \mathbf{O}_{\frac{\pi}{2}}\vec{v} + \vec{w}$, where
$\mathbf{O}_{\frac{\pi}{2}} = \begin{bmatrix} 0 & -1 \\ 1 & 0 \end{bmatrix}$ and $\vec{w} = \begin{bmatrix} 2 \\ 0 \end{bmatrix}$.
For example:
L reflects the origin through the line ℓ to the blue point $\begin{bmatrix} 1 \\ 1 \end{bmatrix}$ and then translates it by the vector $\begin{bmatrix} 1 \\ -1 \end{bmatrix}$ to the point $\vec{w}$.

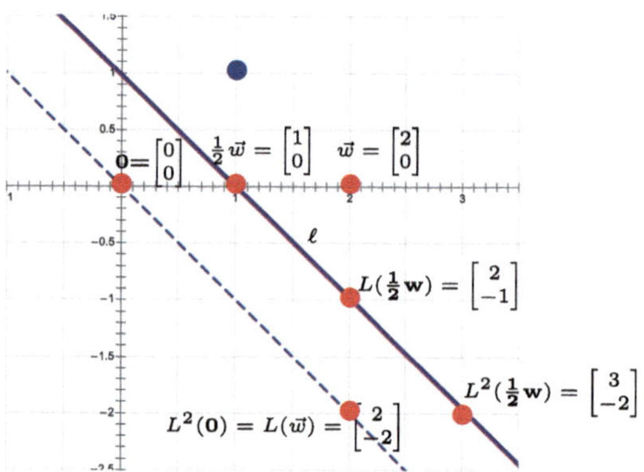

9.2.1 Direct Isometries

Let's delve a little deeper into the direct isometries. According to Theorem 9.2, if L is a direct isometry then it is either a translation or a rotation. Having already considered translations in some depth, we now turn our attention to rotations.

We adopt the notation $R_{[\vec{c},\theta]}$ to denote the counterclockwise rotation by an angle θ about the point $\vec{c}$. Recall that rotations are the only isometries with a unique fixed point. From Theorem 9.2, the general formula for a rotation is:

$$R_{[\vec{c},\theta]}(\vec{v}) = \mathbf{D}_\theta \vec{v} + (\mathbf{I} - \mathbf{D}_\theta)\vec{c}$$

where

$$\mathbf{D}_\theta = \begin{bmatrix} \cos\theta & -\sin\theta \\ \sin\theta & \cos\theta \end{bmatrix}.$$

If the center of rotation $\vec{c}$ is the origin and θ is the angle of rotation, this formula simplifies to:

$$R_{[\vec{0},\theta]}(\vec{v}) = \mathbf{D}_\theta \vec{v}.$$

For example, a rotation of $20°$ about the origin is shown, $R_{[\vec{0},20°]}(\vec{v})$. Each of the three points in the blue triangle are rotated counterclockwise by $20°$. The image is shown in red.

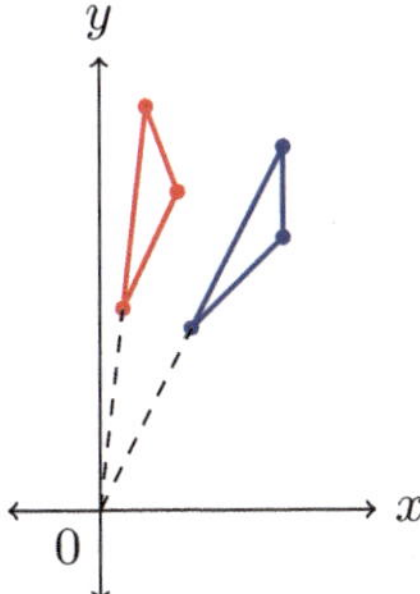

Exercise 9.5 Find the equation of the linear transformation that rotates $180°$ rotation about the origin. Apply this rotation to the point $\vec{p} = \begin{bmatrix} 3 \\ -2 \end{bmatrix}$ and determine the coordinates of $\vec{p}'$.

Exercise 9.6 Apply the rotation $R_{[0,\theta]}$ where $\theta = \frac{\pi}{2}$ to the following points: $\vec{p} = \begin{bmatrix} 1 \\ 2 \end{bmatrix}$ and $\vec{q} = \begin{bmatrix} -3 \\ 1 \end{bmatrix}$.

Exercise 9.7 Write the isometry for $R_{[\vec{0},20°]}(\vec{v})$.

Exercise 9.8 Complete the following table that outlines for the standard rotations between $0°$ and $90°$ in the plane.

θ	*Matrix*	*Unit vector*	*Illustration*
30°		$\mathbf{D}_{30°}\begin{bmatrix} 1 \\ 0 \end{bmatrix} = \begin{bmatrix} \frac{\sqrt{3}}{2} \\ \frac{1}{2} \end{bmatrix}$	
45°	$\mathbf{D}_{45°} = \begin{bmatrix} \frac{1}{\sqrt{2}} & -\frac{1}{\sqrt{2}} \\ \frac{1}{\sqrt{2}} & \frac{1}{\sqrt{2}} \end{bmatrix}$		
60°	$\mathbf{D}_{60°} = \begin{bmatrix} \frac{1}{2} & -\frac{\sqrt{3}}{2} \\ \frac{\sqrt{3}}{2} & \frac{1}{2} \end{bmatrix}$		
90°		$\mathbf{D}_{90°}\begin{bmatrix} 1 \\ 0 \end{bmatrix} = \begin{bmatrix} 0 \\ 1 \end{bmatrix}$	

Now let's shift the center of rotation away from the origin. Earlier in this chapter, we discussed how linear transformations can be composed to create more complex transformations. A rotation about a point other than the origin involves the composition of two linear transformations: a translation and a rotation. In this example, we illustrate a 20° rotation about the point $(-4, 1)$.

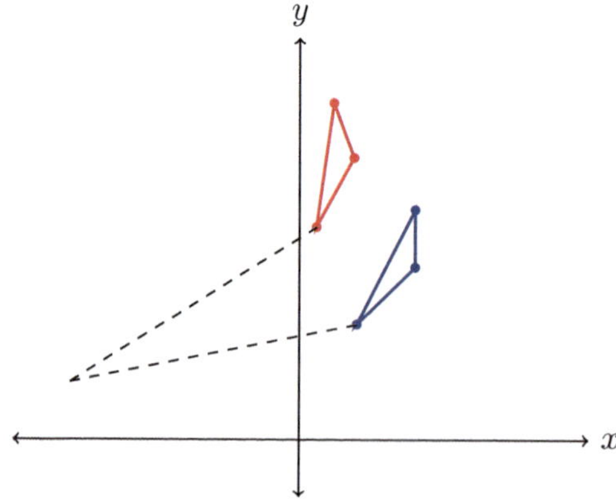

The formula for this rotation is

$$R_{[\vec{c},\theta]}(\vec{v}) = \mathbf{D}_\theta \vec{v} + (\mathbf{I} - \mathbf{D}_\theta)\vec{c}.$$

We now compute this explicitly.

$$R_{[\vec{c},20°]}(\vec{v}) = \begin{bmatrix} \cos 20° & -\sin 20° \\ \sin 20° & \cos 20° \end{bmatrix} \vec{v} + \left(\mathbf{I} - \begin{bmatrix} \cos 20° & -\sin 20° \\ \sin 20° & \cos 20° \end{bmatrix} \right) \begin{bmatrix} -4 \\ 1 \end{bmatrix}$$

$$= \begin{bmatrix} 0.9397 & -0.3420 \\ 0.3420 & 0.9397 \end{bmatrix} \vec{v} + \begin{bmatrix} 0.0603 & 0.3420 \\ -0.3420 & 0.0603 \end{bmatrix} \begin{bmatrix} -4 \\ 1 \end{bmatrix}$$

$$= \begin{bmatrix} 0.9397 & -0.3420 \\ 0.3420 & 0.9397 \end{bmatrix} \vec{v} + \begin{bmatrix} 0.1008 \\ 1.4283 \end{bmatrix}.$$

Exercise 9.9 Confirm that $\vec{v} = \begin{bmatrix} -4 \\ 1 \end{bmatrix}$ is the fixed point of R.

Exercise 9.10 Find the image of a triangle with vertices

$$\vec{a} = \begin{bmatrix} 2 \\ 1 \end{bmatrix}, \quad \vec{b} = \begin{bmatrix} 5 \\ 1 \end{bmatrix}, \quad \vec{c} = \begin{bmatrix} 3 \\ 4 \end{bmatrix},$$

after a rotation about the point $\vec{p} = \begin{bmatrix} -1 \\ -1 \end{bmatrix}$ by 90° counterclockwise.

Exercise 9.11 Find the matrix equation for the rotation by 45° about the point $\vec{p} = \begin{bmatrix} 1 \\ 1 \end{bmatrix}$.

Apply this rotation to the point $\vec{d} = \begin{bmatrix} 3 \\ 2 \end{bmatrix}$ and find the coordinates of the rotated point $\vec{d}'$.

Exercise 9.12 Find the equation of the 90° rotation about the point $\vec{w} = \begin{bmatrix} 2 \\ -3 \end{bmatrix}$. Provide
the transformation matrix and describe how it affects any given point (x, y).

Let's now examine the composition of two rotations and observe the resulting transformation. When both rotations are centered at the same point, it is intuitive to see how their composition affects the plane, as each rotation simply builds on the previous one around the same fixed point.

However, the situation becomes more complex when the rotations have distinct fixed points. To address such cases, the following theorem provides a general formula for the composition of rotations with different centers.

Theorem 9.3 *Consider the rotations $R_{[\mathbf{c},\theta]}$ and $R_{[\mathbf{c}',\phi]}$. If $\theta + \phi = 0$ or 2π, then $R_{[\mathbf{c},\theta]} \circ R_{[\mathbf{c}',\phi]}$ is a translation; otherwise,*

$$R_{[\mathbf{c},\theta]} \circ R_{[\mathbf{c}',\phi]} = R_{[\mathbf{c}'',\,\theta+\phi]}$$

for some point $\mathbf{c}''$.

Proof Recall that the rotation by angle θ about the point $\vec{c}$ is given by

$$R_{[\vec{c},\theta]}(\vec{v}) = \mathbf{D}_\theta \vec{v} + (\mathbf{I} - \mathbf{D}_\theta)\vec{c} \quad where \quad \mathbf{D}_\theta = \begin{bmatrix} \cos\theta & -\sin\theta \\ \sin\theta & \cos\theta \end{bmatrix}.$$

Similarly,

$$R_{[\vec{c}',\phi]}(\vec{v}) = \mathbf{D}_\phi \vec{v} + (\mathbf{I} - \mathbf{D}_\phi)\vec{c}' \quad where \quad \mathbf{D}_\phi = \begin{bmatrix} \cos\phi & -\sin\phi \\ \sin\phi & \cos\phi \end{bmatrix}.$$

We compute the composition:

$$R_{[\vec{c},\theta]} \circ R_{[\vec{c}',\phi]}(\vec{v}) = \mathbf{D}_\theta \left(\mathbf{D}_\phi \vec{v} + (\mathbf{I} - \mathbf{D}_\phi)\vec{c}' \right) + (\mathbf{I} - \mathbf{D}_\theta)\vec{c}$$

$$= \mathbf{D}_\theta \mathbf{D}_\phi \vec{v} + \mathbf{D}_\theta (\mathbf{I} - \mathbf{D}_\phi)\vec{c}' + (\mathbf{I} - \mathbf{D}_\theta)\vec{c}.$$

Let

$$\vec{u} = \mathbf{D}_\theta (\mathbf{I} - \mathbf{D}_\phi)\vec{c}' + (\mathbf{I} - \mathbf{D}_\theta)\vec{c}.$$

Then

$$R_{[\vec{c},\theta]} \circ R_{[\vec{c}',\phi]}(\vec{v}) = \mathbf{D}_\theta \mathbf{D}_\phi \vec{v} + \vec{u}.$$

The key to this composition is the matrix product $\mathbf{D}_\phi \mathbf{D}_\theta = \mathbf{D}_\theta \mathbf{D}_\phi = \mathbf{D}_{\theta+\phi}$,

$$\mathbf{D}_\phi \mathbf{D}_\theta = \begin{bmatrix} \cos\phi & -\sin\phi \\ \sin\phi & \cos\phi \end{bmatrix} \begin{bmatrix} \cos\theta & -\sin\theta \\ \sin\theta & \cos\theta \end{bmatrix}$$

$$= \begin{bmatrix} \cos\phi\cos\theta - \sin\phi\sin\theta & -(\cos\phi\sin\theta + \sin\phi\cos\theta) \\ \sin\phi\cos\theta + \cos\phi\sin\theta & \cos\phi\cos\theta - \sin\phi\sin\theta \end{bmatrix}$$

$$= \begin{bmatrix} \cos(\theta+\phi) & -\sin(\theta+\phi) \\ \sin(\theta+\phi) & \cos(\theta+\phi) \end{bmatrix}.$$

Since rotation matrices commute and add their angles,

$$\mathbf{D}_\theta \mathbf{D}_\phi = \mathbf{D}_{\theta+\phi}.$$

If $\theta + \phi \neq 0, 2\pi$, then $\mathbf{I} - \mathbf{D}_{\theta+\phi}$ is invertible, so there exists a point $\vec{c}''$ such that

$$\vec{u} = (\mathbf{I} - \mathbf{D}_{\theta+\phi})\vec{c}''.$$

Thus,

$$R_{[\vec{c},\theta]} \circ R_{[\vec{c}',\phi]}(\vec{v}) = \mathbf{D}_{\theta+\phi}\vec{v} + (\mathbf{I} - \mathbf{D}_{\theta+\phi})\vec{c}'' = R_{[\vec{c}'',\,\theta+\phi]}(\vec{v}).$$

If $\theta + \phi = 0$ or 2π, then $\mathbf{D}_{\theta+\phi} = \mathbf{I}$, and the composition reduces to

$$R_{[\vec{c},\theta]} \circ R_{[\vec{c}',\phi]}(\vec{v}) = \vec{v} + \vec{u},$$

which is a translation.   □

Corollary 9.1 *Consider two rotations $R_{[c,\theta]}$ and $R_{[c,\phi]}$ with the same center c. Then $R_{[c,\theta]} \circ R_{[c,\phi]} = R_{[c,\theta+\phi]}$. In the special case $\theta + \phi = 0$ or 2π, $R_{[c,2\pi]} = R_{[c,0]}$ is the identity transformation.*

Exercise 9.13 Prove Corollary 9.1

Exercise 9.14 Compute the following composition of isometries and describe each of them geometrically.

a. $R_{[\vec{0},30°]} \circ R_{[\vec{0},60°]}$

b. $R_{[\vec{0},60°]} \circ R_{[\vec{0},60°]}$

c. $R_{[\vec{0},90°]} \circ R_{[\vec{0},30°]}$

Exercise 9.15 Find the equation of the 90° rotation about the point $\vec{p} = \begin{bmatrix} 2 \\ -3 \end{bmatrix}$ followed by a rotation of 30° around the origin.

Provide the transformation matrix and describe how it affects any given point $\vec{v} = \begin{bmatrix} x \\ y \end{bmatrix}$.

We have discussed how to determine the formula for a rotation based on its angle and center of rotation. Next, consider a scenario in which we are given two figures and aim to write the formula for the rotation using the image alone. While it may be straightforward to recognize that the transformation is a rotation, identifying the exact angle of rotation is less obvious. Furthermore, we must use constructions to determine the exact center of rotation which cannot be done by eyesight alone.

To identify the formula for a rotation from an image, we start by selecting at least two pairs of corresponding points between the pre-image and its image. Suppose point A in the pre-image maps to point A' in the image, and point B maps to point B' as shown below. These pairs of points serve as the foundation for identifying both the center of rotation and the angle of rotation.

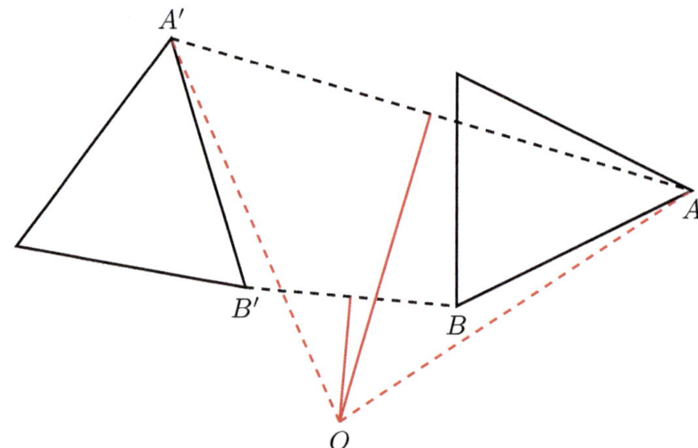

To determine the angle of rotation, begin by constructing the perpendicular bisectors of the segments $\overline{AA'}$ and $\overline{BB'}$, shown as the solid red lines in the figure. The point where these two bisectors intersect is the center of rotation, denoted as O. This point is equidistant from both A and A', as well as from B and B'.

Once the center of rotation is identified, draw the segments $\overline{OA}$ and $\overline{OA'}$ (shown as dashed red lines in the figure). Measure the angle $\angle AOA'$ using a protractor to find the angle of rotation. Be sure to note whether the rotation is clockwise or counterclockwise. To verify the rotation, check another pair of corresponding points, such as B and B', ensuring that the angle $\angle BOB'$ matches $\angle AOA'$ and that the direction of rotation is consistent.

Exercise 9.16 Consider the five congruent triangles labeled Y, B, R, P, and G in Fig. 9.1 In Exercise 9.2 you identified those pairs of triangles related by direct isometries.

a. Of the triangles related by direct isometries determine which pairs are related by translation and which are related by rotation?
b. For each pair, identify a rotation or translation that maps one of them onto the other.

We conclude this section with an application of rotations to solve a classical optimization problem known as Fermat's Problem. The goal of Fermat's Problem is to determine the point inside a triangle that minimizes the total distance to the three vertices of the triangle. This point is known as the Fermat Point of the triangle.

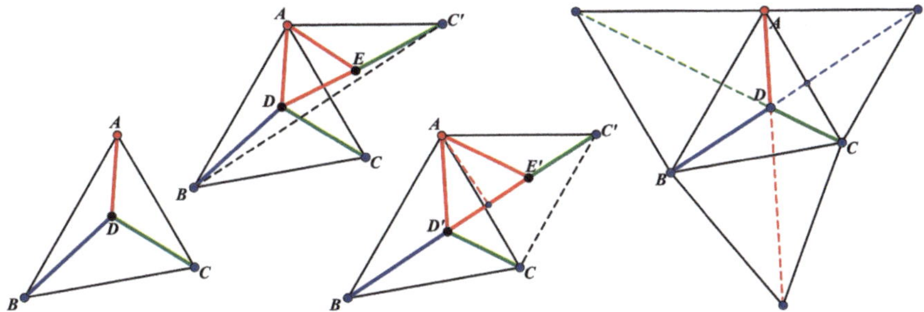

Consider the triangle $\triangle ABC$ in the lower-left of the figure, and let D represent a potential position for the Fermat Point. To verify if D is indeed the Fermat Point, we rotate $\triangle ADC$ 60° counterclockwise about vertex A. This rotation is shown in the second diagram of the figure, where E marks the image of D under the rotation.

We show that $\triangle ADE$ forms an equilateral triangle Since segments $\overline{AD}$ and $\overline{AE}$ have equal lengths, $\triangle ADE$ is isosceles. Furthermore, the angles $\angle ADE$ and $\angle DEA$ must be equal and sum to 120°, implying they are each 60°. Consequently, DE also has the same length as AD. Since EC' is the image of $\overline{DC}$ under a rotation, $\overline{EC'}$ has the same length as $\overline{DC}$.

Thus, the sum of the distance from D to the vertices A, B, and C equals the length of the polygonal line $BDEC'$. Additionally, we observe that the length of segment $\overline{BC'}$ represents the minimum possible sum of distances from any interior point to the triangle's vertices.

Now can we choose the position of D along $\overline{BC'}$ so that its rotated image E also lies on $\overline{BC'}$? The answer is yes: by dropping a perpendicular from A to segment $\overline{BC'}$, we can construct two 30-60-90 right triangles, as shown in the third diagram. The point D', constructed this way is the Fermat Point.

Finally, note that $\triangle ACC'$ is also equilateral: again $\overline{AC}$ and $\overline{AC'}$ have equal lengths and $\angle CAC' = 60°$, so $\triangle ACC'$ must be equilateral. Moreover, the segment connecting vertex

B to the third vertex of the external equilateral triangle constructed on the opposite side of $\overline{AC}$ passes through D'. By symmetry, this argument holds for any vertex and its opposite external equilateral triangle.

Hence, the segments connecting each vertex to the corresponding vertex of the opposite external equilateral triangle are concurrent at the Fermat Point, and their lengths are equal.

9.2.2 Opposite Isometries

Let's shift our focus from direct isometries to opposite isometries. Unlike direct isometries, which preserve the original orientation, opposite isometries—such as reflections and glide reflections—reverse the figure's orientation by flipping it across a line of reflection. Let's begin by examining basic reflections and later explore glide reflections, which combine a reflection with a translation.

Deriving the formula for an opposite isometry is more complex than determining the formulas for direct isometries. In the previous theorems, we started with a given isometry and used its properties to identify the axis of reflection or the associated translation. Now, we approach the problem from the reverse direction: how to determine the formula for an opposite isometry when the explicit formula is unknown, but rather we only have the direction vector $\vec{d}$ of the reflection axis, ℓ.

If we have only a figure to work with, we can use constructions to determine the axis of reflection for an opposite isometry. To do so, we select two pairs of corresponding points: $\{A, A'\}$ and $\{B, B'\}$. The axis of reflection is the line that passes through the midpoints of $\overline{AA'}$ and $\overline{BB'}$, as shown in red in the figure below.

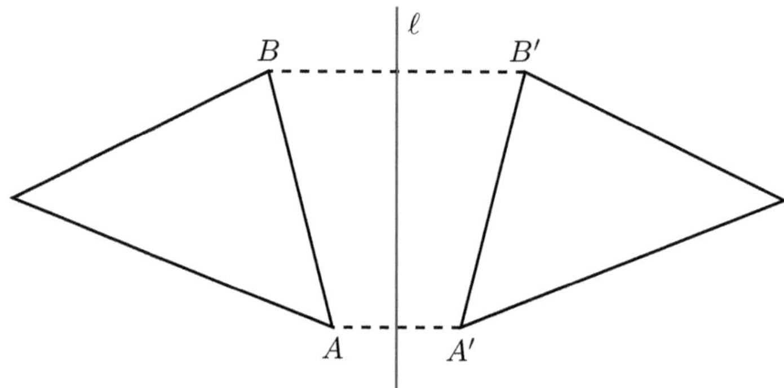

Exercise 9.17 Find axis of reflection for the triangles in Fig. 9.1 that are paired by an opposite isometry.

Once the line of reflection for a transformation has been identified, the first step in writing the isometry is to parametrize the line, ℓ. Let $\ell = \lambda \vec{d} + \vec{u}$ where $\vec{d}$ is the direction vector and $\vec{u}$ is any point on ℓ. Since we can choose $\vec{u}$ to be any vector with an endpoint on ℓ, we choose it to be perpendicular to the direction vector $\vec{d}$. Next, let θ be twice the counterclockwise angle that $\vec{d}$ makes with the positive x-axis. Using this angle, the transformation is given by the formula $L(\vec{v}) = \mathbf{O}_\theta \vec{v} + \vec{w}$, where $\mathbf{O}_\theta$ is the reflection matrix for the angle θ. Furthermore, we must determine the appropriate vector $\vec{w}$ based on $\vec{d}$ and $\vec{u}$.

For a simple reflection across the line ℓ, the vector $\vec{w}$ is defined as $2\vec{u}$. Since $\vec{u}$ was chosen to be perpendicular to $\vec{d}$, $\vec{w}$ must be too. Therefore the reflection matrix $\mathbf{O}_\theta$ reflects $\vec{w}$ through the line $\lambda \vec{d}$. Equivalently this can be expressed as: $\mathbf{O}_\theta \vec{w} = -\vec{w}$. It follows that $\mathbf{O}_\theta \vec{w} + \vec{w} = 0$ when $\vec{w} = 2\vec{u}$, which confirms that the isometry $L(\vec{v}) = \mathbf{O}_\theta \vec{v} + 2\vec{u}$ represents the reflection across the line $\lambda \vec{d} + \vec{u}$.

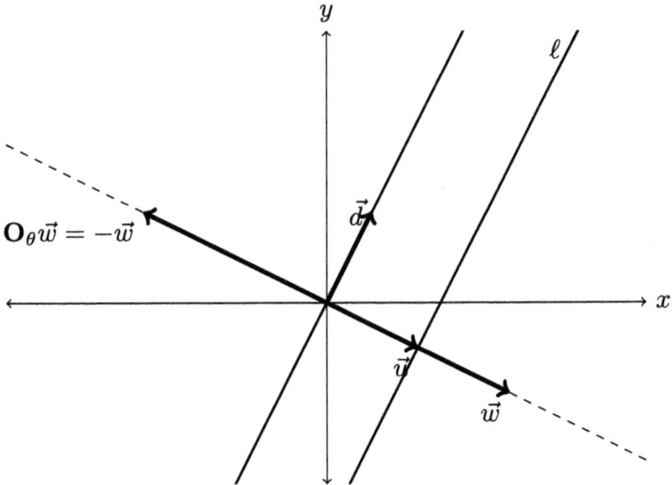

Now let's consider a glide reflection. In this case, we aim to determine the formula for the isometry based on the axis of reflection, $\ell = \lambda \vec{d} + \vec{u}$, and the associated translation. The translation must be chosen in the same direction as ℓ, so it has the form $v\mathbf{d}$. Notice that the vector $\vec{u}' = \vec{u} + v\vec{d}$ still lies on the line ℓ. As such, ℓ can also be written parametrically as $\lambda \vec{d} + \vec{u}'$. Now, if we define $\vec{w}' = 2\vec{u}'$ then the isometry becomes $L(\vec{v}) = \mathbf{O}_\theta \vec{v} + \vec{w}'$.

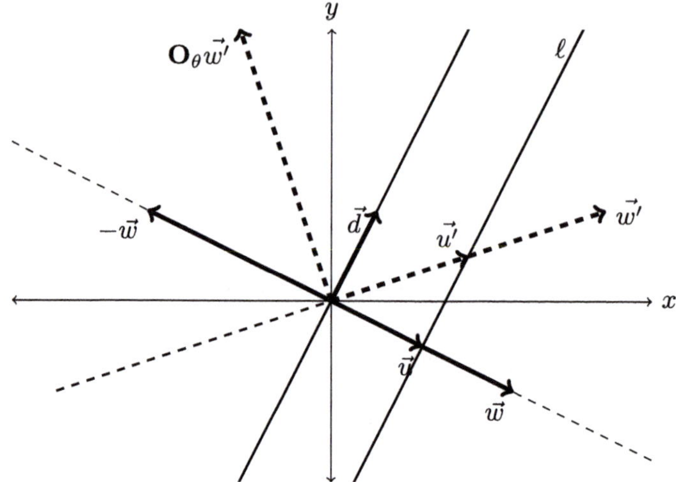

However, due to the translation, the direction of $\vec{u}'$ is no longer orthogonal to the direction of ℓ. It follows that the reflected vector $\mathbf{O}_\theta \vec{w}'$ is no longer equal to $-\vec{w}'$. The vectors $\vec{w}'$ and $\mathbf{O}_\theta \vec{w}'$ are illustrated in the figure above. Since $\mathbf{O}_\theta \vec{w}' \neq -\vec{w}$, it clearly follows that $\mathbf{O}_\theta \vec{w}' + \vec{w}' \neq 0$, so by Theorem 9.2, we have that $L(\vec{v}) = \mathbf{O}_\theta \vec{v} + \vec{w}'$ is a glide reflection. We can substitute in $\vec{w}' = 2\vec{u}' = 2(\vec{u} + v\vec{d})$

$$L(\vec{v}) = \mathbf{O}_\theta \vec{v} + 2(\vec{u} + v\vec{d}),$$

as the glide reflection across the line $\lambda\vec{d} + \vec{u}$ with a translation of $v\vec{d}$.

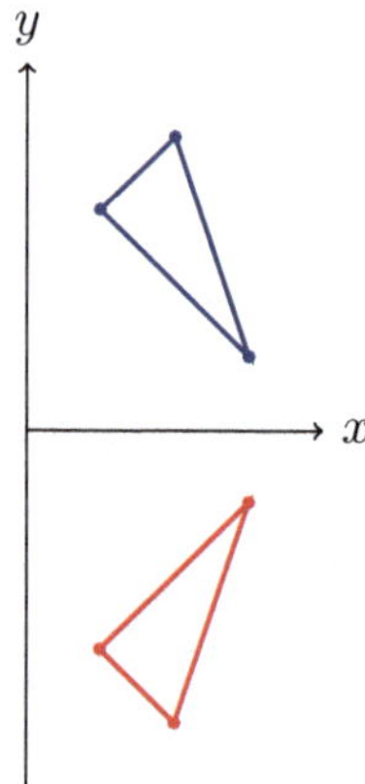

As an example, let's consider a simple reflection of the the form $M(\vec{v}) = \mathbf{A}\vec{v}$, where $\mathbf{A}$ is a reflection matrix. In this case, the reflection matrix is given by

$$A = \begin{bmatrix} 1 & 0 \\ 0 & -1 \end{bmatrix}$$

Note that the matrix A leaves the x-coordinate unchanged while negating the y-coordinate. Therefore, this reflection flips the figure over the x-axis as shown in the figure below.

Exercise 9.18 Find the isometry for the reflection across the line $x = -1$. How does this change if we add a translation by $\vec{t} = \begin{bmatrix} 1 \\ 1 \end{bmatrix}$?

Exercise 9.19 Determine the isometry for reflecting across the line $x = 2$. Compute the new coordinates of the point $\vec{p} = \begin{bmatrix} 5 \\ -3 \end{bmatrix}$ after reflection.

Exercise 9.20 Consider a glide reflection with an axis of reflection along $y = 3$ and a translation vector $\vec{v} = \begin{bmatrix} 2 \\ 0 \end{bmatrix}$.

a. Graph the image of a parallelogram with vertices

$$\vec{a} = \begin{bmatrix} 1 \\ 2 \end{bmatrix}, \quad \vec{b} = \begin{bmatrix} 4 \\ 2 \end{bmatrix}, \quad \vec{c} = \begin{bmatrix} 4 \\ 5 \end{bmatrix}, \quad \vec{d} = \begin{bmatrix} 1 \\ 5 \end{bmatrix}.$$

b. Find the isometry corresponding to this glide reflection. Show your work.

   As discussed earlier, a glide reflection can be understood as a two-step transformation: first, a figure is reflected across a line, and then it is translated along a vector. This transformation can be expressed as the composition: $L(\vec{v}) = T_{\vec{w}}(K(\vec{v}))$, where K represents the reflection and $T_{\vec{w}}$ represents the translation by the vector $\vec{w}$. However we know we can also write this transformation as $L(\vec{v}) = O_\theta \vec{v} + 2(\vec{u} + v\vec{d})$ where our line of reflection has direction $\vec{d}$, θ is the angle between $\vec{d}$ and the positive x-axis and our translation vector is $v\vec{d}$

   Let's consider the glide reflection shown below.

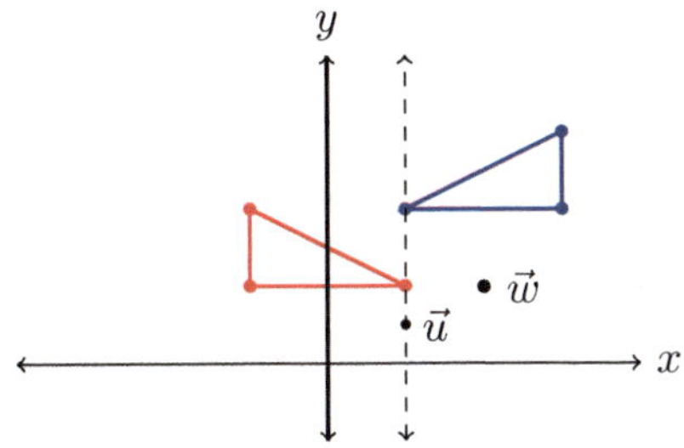

In this example, the blue triangle with vertices, $\begin{bmatrix} -1 \\ 1 \end{bmatrix}$, $\begin{bmatrix} -1 \\ 2 \end{bmatrix}$ and $\begin{bmatrix} 1 \\ 1 \end{bmatrix}$, can be reflected across the y-axis and then translated by vector $\vec{w} = \begin{bmatrix} 2 \\ 1 \end{bmatrix}$. This results in the red triangle with vertices, $\begin{bmatrix} 1 \\ 2 \end{bmatrix}$, $\begin{bmatrix} 3 \\ 2 \end{bmatrix}$, and $\begin{bmatrix} 3 \\ 3 \end{bmatrix}$. However, the y-axis is not fixed by this set of isometries so is not considered the axis of reflection.

Since the reflection across the y-axis leaves the y-coordinate fixed while negating the x-coordinate, we use the reflection matrix:

$$K = \begin{bmatrix} -1 & 0 \\ 0 & 1 \end{bmatrix},$$

Adding the translation $\vec{w} = \begin{bmatrix} 2 \\ 1 \end{bmatrix}$, the glide reflection can be represented as:

$$L(\vec{v}) = \begin{bmatrix} -1 & 0 \\ 0 & 1 \end{bmatrix} \vec{v} + \begin{bmatrix} 2 \\ 1 \end{bmatrix}.$$

To identify the axis of reflection, we recall that that the point

$$\vec{u} = \frac{1}{2}\vec{w} = \begin{bmatrix} 1 \\ \frac{1}{2} \end{bmatrix}$$

will lie on the axis of reflection which we can easily see must be parallel to the y-axis. It follows that the line of reflection is $x = 1$ and we can compute the translation vector to be

$$\frac{K(\vec{w}) + \vec{w}}{2} = \frac{1}{2}(\begin{bmatrix} -1 & 0 \\ 0 & 1 \end{bmatrix}\begin{bmatrix} 2 \\ 1 \end{bmatrix} + \begin{bmatrix} 2 \\ 1 \end{bmatrix}) = \begin{bmatrix} 0 \\ 1 \end{bmatrix}$$

So we have the axis of reflection that is fixed by this transformation is $x = 1$ and the associated translation is $\begin{bmatrix} 0 \\ 1 \end{bmatrix}$.

Exercise 9.21 Find the isometry for a glide reflection with axis of reflection $y = -2$ and translation vector $\vec{v} = \begin{bmatrix} -4 \\ 2 \end{bmatrix}$.

Apply this glide reflection to the point $\vec{p} = \begin{bmatrix} 3 \\ 1 \end{bmatrix}$ and find the coordinates of the transformed point $\vec{p}'$.

Exercise 9.22 Consider $L(\vec{v}) = \mathbf{O}_\theta \vec{v} + \vec{w}$ where $\vec{w} = \begin{bmatrix} 4 \\ -1 \end{bmatrix}$ and $\theta = \frac{\pi}{2}$. Give a precise description this transformation, including its two-grid graph.

Now, let's return to our five triangles from previous exercises.

Exercise 9.23 Recall the five congruent triangles labeled Y, B, R, P, and G in Fig. 9.1 In Exercise 9.2 you identified those pairs of triangles related by opposite isometries.

a. Of the triangles related by opposite isometries determine which pairs are related by a reflection and which are related by a glide reflection?
b. For each pair, identify a rotation or translation that maps one of them onto the other.

Exercise 9.24 Consider the five triangles above. Recall, each of these triangles is an isometry of the others. Now, find the axis of reflection for the pairs of triangles that are related by reflection.

We now turn our attention to Fagnano's Problem. Just as Fermat's Problem showcased the intriguing applications of rotations in geometry, Fagnano's Problem illustrates the power and elegance of reflections in geometric problem-solving. The problem is stated as follows: given an acute-angled triangle $\triangle(\vec{a}, \vec{b}, \vec{c})$, find the inscribed triangle with the minimum perimeter. In the diagram on the left below, we illustrate the triangle $\triangle(\vec{a}, \vec{b}, \vec{c})$ alongside an initial freehand attempt to identify the inscribed triangle of minimum perimeter. This initial attempt is labeled as triangle $\triangle(\vec{d}, \vec{e}, \vec{f})$ in the figure.

Our first question is: How accurate is our initial approximation? To explore this, we constructed the second diagram. In this diagram, the triangle $\triangle(\vec{g}, \vec{a}, \vec{h})$ is formed by reflecting the segment $\vec{a}\vec{d}$, first across the side $\vec{a}\vec{c}$ and then across the side $\vec{a}\vec{b}$. Clearly, the perimeter of $\triangle(\vec{d}, \vec{e}, \vec{f})$ can be shortened by adjusting $\vec{e}$ to the intersection of the segment $\vec{g}\vec{h}$ with the left side of $\triangle(\vec{a}, \vec{b}, \vec{c})$ and moving $\vec{f}$ to the intersection of $\vec{g}\vec{h}$ with the right side of $\triangle(\vec{a}, \vec{b}, \vec{c})$. This modification is illustrated in the third diagram.

From this construction, we can conclude that this is the inscribed triangle with the shortest perimeter that passes through the point $\vec{d}$.

This, however, raises a second question: How should we determine the optimal position for $\vec{d}$?

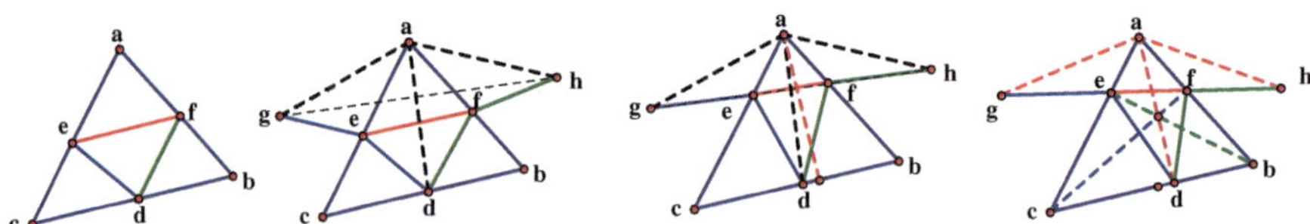

To address this question, we return to the second diagram and examine the triangle $\triangle(\vec{g}, \vec{a}, \vec{h})$. We observe that it is isosceles, with the two equal sides having the same length as $\vec{ad}$, and its apex angle being twice the angle $\angle(\vec{c}, \vec{a}, \vec{b})$. Notably, regardless of the position of $\vec{d}$, the apex angle remains unchanged. Consequently, all triangles $\triangle(\vec{g}, \vec{a}, \vec{h})$ are similar, no matter the choice of $\vec{d}$. Therefore, to minimize the length of the base $\vec{gh}$, we minimize the length of the equal sides by selecting $\vec{ad}$ as the altitude from $\vec{a}$.

In the final diagram, we reposition $\vec{d}$ to the foot of the altitude from $\vec{a}$ and adjust $\vec{e}$ and $\vec{f}$ to lie on the base of the new triangle $\triangle(\vec{g}, \vec{a}, \vec{h})$. By symmetry, the new position of $\vec{e}$ is the foot of the altitude from $\vec{b}$, and the new position of $\vec{f}$ is the foot of the altitude from $\vec{c}$. The concurrency of the altitudes of a triangle is a well-established theorem. For our discussion, we use the result without providing a proof.

9.3   Three Reflection Theorem

The fact that any isometry of a line can be expressed as the composition of at most two reflections highlights a key property of linear functions. Whether a transformation represents translations, reflections, or rotations, it can always be decomposed into a sequence of two or fewer reflections. Extending this principle to transformations in the plane leads to the broader and more significant result known as the Three Reflection Theorem. This theorem states that any isometry of the plane, no matter how complex, can be constructed as the composition of three or fewer reflections.

Theorem 9.4 *Any isometry L of the Euclidean plane can be written as the composition of three or fewer reflections.*

Proof Let L be an isometry and select any set of three non-collinear points. In the accompanying diagram, we use the vertices of the triangle $\triangle(\vec{a}, \vec{b}, \vec{c})$ as our chosen points. The image of this triangle under L is the congruent triangle $\triangle(\vec{a}^*, \vec{b}^*, \vec{c}^*)$.

To show that L can be expressed as the composition of at most three reflections, we aim to find three reflections K_1, K_2, K_3 such that their composition, denoted H, satisfies $H(\vec{a}) = \vec{a}^*$, $H(\vec{b}) = \vec{b}^*$, and $H(\vec{c}) = \vec{c}^*$. By the Three Point Theorem (Theorem 6.2), this would imply that $L = H$.

If $\vec{a} = \vec{a}^*$, $\vec{b} = \vec{b}^*$, and $\vec{c} = \vec{c}^*$, then L is the identity transformation, which can be thought of as either the composition of zero reflections or as the composition of two identical reflections.

Now, assume $\vec{a} \neq \vec{a}^*$ while $\vec{b} = \vec{b}^*$ and $\vec{c} = \vec{c}^*$. Since $\triangle(\vec{a}, \vec{b}, \vec{c})$ and $\triangle(\vec{a}, \vec{b}, \vec{c})$ are congruent, one of the first two configurations in the figure must hold.

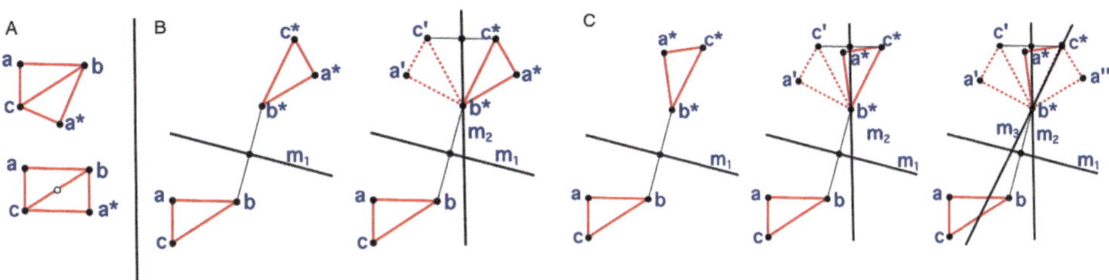

In the top configuration, L is the reflection through the line containing the segment $\overline{bc}$. In the bottom configuration, L is a 180° rotation about the midpoint of the segment $\overline{bc}$, which can also be described as the composition of the reflection through the line containing $\overline{bc}$ and the reflection through the line perpendicular to $\overline{bc}$ at its midpoint.

We leave as an exercise the cases that $\vec{a} \neq \vec{a}^*$, $\vec{b} \neq \vec{b}^*$, but $\vec{c} = \vec{c}^*$ and turn to the cases where $\vec{a} \neq \vec{a}^*$, $\vec{b} \neq \vec{b}^*$ and $\vec{c} \neq \vec{c}^*$.

Consider the first configuration in each of Grouping B and Grouping C. In Grouping B, we observe that $\triangle(\vec{a}, \vec{b}, \vec{c})$ and $\triangle(\vec{a}^*, \vec{b}^*, \vec{c}^*)$ have the same orientation. Hence, we expect L to be the composition of two reflections. In Grouping C, we observe that $\triangle(\vec{a}, \vec{b}, \vec{c})$ and $\triangle(\vec{a}^*, \vec{b}^*, \vec{c}^*)$ have opposite orientations. Therefore, we expect L to be the composition of one or three reflections.

Turning to Grouping B first, we construct the perpendicular bisector of the $\vec{b}, \vec{b}^*$ segment, denoted as line $\vec{m}_S$. The reflection of $\triangle abc$, labeled $\triangle \vec{a}'\vec{b}^*\vec{c}'$, appears in the second configuration of Grouping B. Here, we add the $\overline{c'c^*}$ segment and its perpendicular bisector, $\vec{m}_2$. Since $\triangle \vec{b}^*\vec{c}'\vec{c}^*$ is isosceles, $\vec{m}_2$ passes through $\vec{b}^*$. Finally, because $\triangle \vec{b}^*\vec{c}'\vec{c}^*$ and $\triangle \vec{a}^*\vec{b}^*\vec{c}^*$ are congruent, the reflection through $\vec{m}_2$ must also map $\vec{a}'$ onto $\vec{a}^*$. Hence, L is again the composition of two reflections.

Turning to Grouping C, we similarly construct the axis of reflection $\vec{m}_1$ and, in the second configuration, include the reflection of $\triangle a\vec{b}c$, labeled $\triangle \vec{a}'\vec{b}^*\vec{c}'$. Again, we add the $\overline{c'c^*}$ segment and its perpendicular bisector, $\vec{m}_2$. In this case, the reflection of $\triangle \vec{a}'\vec{b}^*\vec{c}'$ is $\triangle \vec{a}''\vec{b}^*\vec{c}^*$. To map $\triangle \vec{a}''\vec{b}^*\vec{c}^*$ onto $\triangle \vec{a}^*\vec{b}^*\vec{c}^*$, we need a reflection through line $\vec{m}_3$, which contains the vertices $\vec{b}^*$ and $\vec{c}^*$. Thus, in this case, L is the composition of three reflections.

$\square$

When composing two reflections, the resulting transformation depends on the orientation of their axes. If the axes are parallel, the composition is a translation, with the translation vector being perpendicular to the axes and twice the distance between them. If the axes intersect, the composition results in a rotation around the intersection point, with an angle that is twice the angle between the axes. The following theorem formalizes this result.

Theorem 9.5 *Let R and S be reflections with axes m_R and m_S, respectively.*

(1) If m_R and m_S are parallel then $S \circ R$ is the translation by the vector perpendicular to m_R, directed from m_R toward m_S with length twice the distance between m_R and m_S

(2) If m_R and m_S intersect at $\vec{c}$, then $S \circ R$ is the counterclockwise rotation about $\vec{c}$ by twice the counterclockwise angle from m_R toward m_S.

Proof Begin by considering the case where m_R and m_S are parallel lines with a distance d between them. Consider an arbitrary point $\vec{p}$ in the plane. The image $R(\vec{p})$ will lie on the line perpendicular to m_R that passes through $\vec{p}$. $R(\vec{p})$ lies on the opposite side of m_R at the same distance as $\vec{p}$ from m_R. Shown using the red dashed lines.

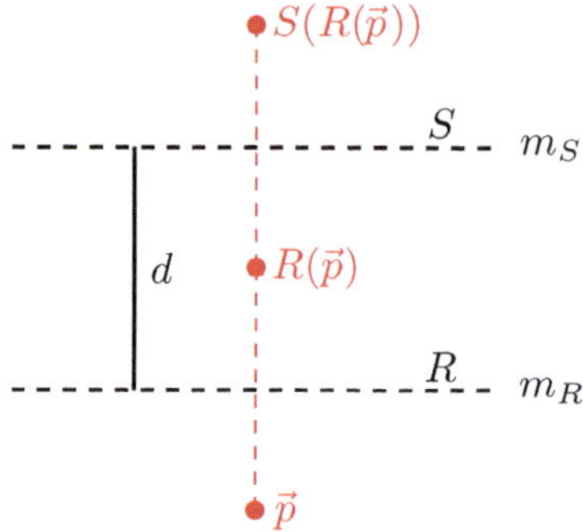

Now the image of $R(\vec{p})$ after reflecting across m_S, lies on the opposite side of m_S on the line perpendicular to m_S that passes through $R(\vec{p})$. $S(R(\vec{p}))$ and $R(\vec{p})$ are equidistance from m_S. Since m_R and m_S are parallel, the line passing through $\vec{p}$, $R(\vec{p})$, and $S(R(\vec{p}))$ is perpendicular to both axes. So we see that the composition of these two reflections is a translation by a vector perpendicular to the axes, directed from m_R towards m_S, with length equal to twice the distance d between the axes.

Now consider the case where m_R and m_S intersect at a point $\vec{c}$, and let θ be the angle from m_R to m_S, measured counterclockwise.

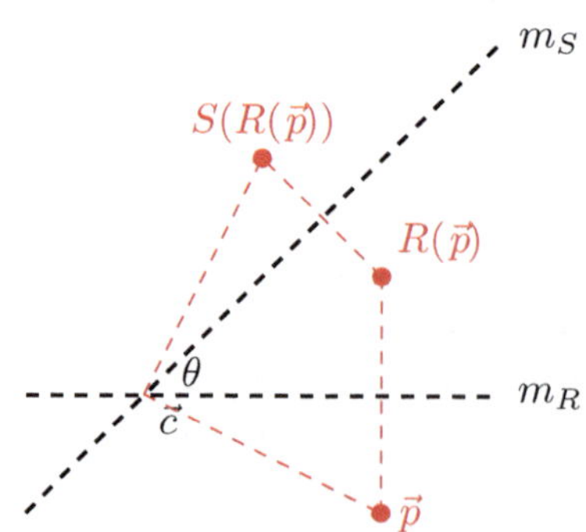

After reflecting our point $\vec{p}$ across m_R, the image $R(\vec{p})$ lies on the opposite side of m_R, with m_R forming the perpendicular bisector of the segment $\vec{p}R(\vec{p})$. Reflecting $R(\vec{p})$ across m_S, the image $S(R(\vec{p}))$ lies on the opposite side of m_S, with m_S as the perpendicular bisector of the segment $R(\vec{p})S(R(\vec{p}))$. We see the composition $S \circ R$ is equivalent to rotating $\vec{p}$ about $\vec{c}$ by an angle 2θ, where θ is the angle between m_R and m_S. The direction of the rotation is counterclockwise if θ is measured counterclockwise from m_R to m_S. $\square$

Although Theorem 9.4 tells us that all isometries can be represented by three reflections, the following theorem specifically demonstrates the method to decompose any glide reflection into three reflections. First, reflect across the axis of reflection to capture the matrix component of the transformation. Then, perform two additional reflections: one across a line parallel to the axis and another across a line perpendicular to it, which together produce the translation component of the glide reflection. The following theorem formalizes this decomposition.

Theorem 9.6 *Let G be a glide reflection with axis of reflection ℓ and translation vector $\vec{w}$. Then G can be written as a composition of three reflections:*

(1) A reflection R across the axis ℓ.
(2) A reflection S across a line m_S perpendicular to ℓ.
(3) A reflection T across a line m_T parallel to m_S, with m_T located such that $T \circ S$ produces the translation by $\vec{w}$.

Thus, the glide reflection G can be expressed as $G = T \circ S \circ R$.

Proof To prove that any glide reflection can be decomposed into three reflections, we begin by noting that a glide reflection G is composed of a reflection across an axis ℓ followed by a translation along a vector $\vec{w}$ parallel to ℓ. Let this translation be denoted as $T_{\vec{w}}$.

From the first part of Theorem 9.6, we know that the translation $T_{\vec{w}}$ can be expressed as the composition of two reflections across parallel lines perpendicular to ℓ. Specifically, let m_S be a line parallel to ℓ. Reflecting across m_S, followed by a reflection across another parallel line m_T, also perpendicular to ℓ, results in $T_{\vec{w}} = T \circ S$. Therefore, the glide reflection G can be expressed as

$$G = T_{\vec{w}} \circ R = (T \circ S) \circ R.$$

$\square$

10.1 Symmetry Groups of Polygons

A regular polygon with n sides has an associated symmetry group that includes both rotations and reflections. This symmetry group consists of all isometries that map the polygon onto itself, preserving its shape and orientation. By positioning a regular polygon with its center at the origin, these isometries are all matrix isometries.

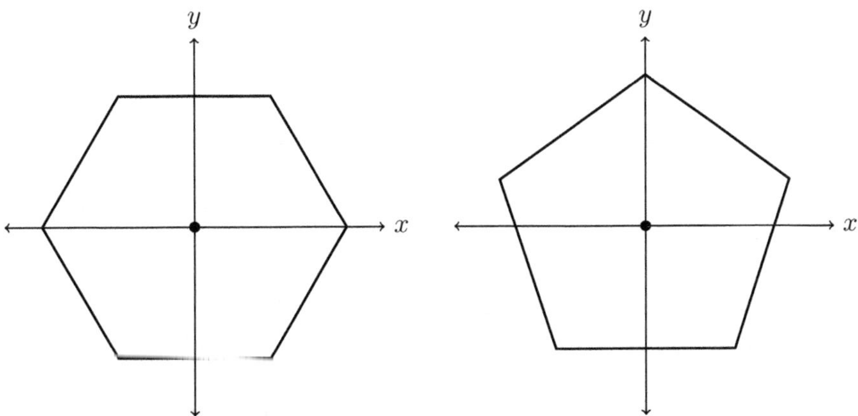

A regular n-sided polygon can be rotated about its center by angles that are multiples of $\frac{360°}{n}$, with each rotation being a matrix isometry of the form $L(\vec{v}) = \mathbf{D}_{\frac{360°}{n}}\,\vec{v}$. For example, a regular hexagon ($n = 6$) has rotational symmetries at $60°$, $120°$, $180°$, $240°$, and $300°$, in addition to the identity rotation ($0°$), as illustrated in the figure below. Similarly, a regular pentagon ($n = 5$) exhibits rotational symmetries at $0°$, $72°$, $144°$, $216°$, and $288°$. These rotational symmetries are highlighted with red dashed lines in the accompanying figure.

© The Author(s), under exclusive license to Springer Nature Switzerland AG 2026

J. J. Edmond, J. E. Graver, *The Linear Function and Euclidean Geometry*, Synthesis Lectures on Mathematics & Statistics, https://doi.org/10.1007/978-3-032-22712-6_10

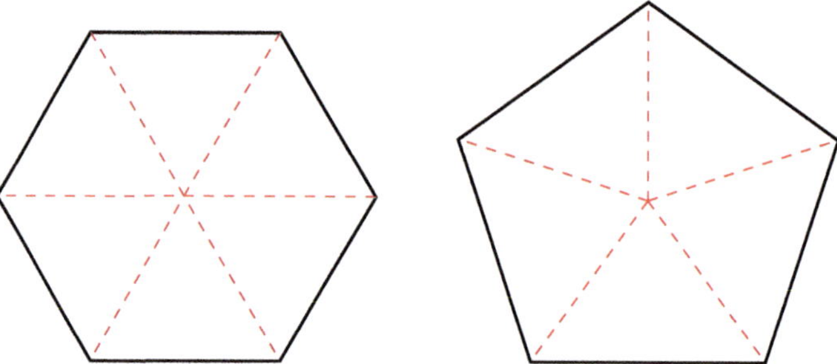

A regular n-sided polygon also has n axes of reflection. For a hexagon, reflecting across any of the dashed lines from the rotational symmetry figure above will map the hexagon onto itself. Since the hexagon has an even number of sides, these reflection lines pass through pairs of opposite vertices. Additionally, there is another set of three axes of reflection, shown in blue; these axes pass through the midpoints of opposite sides. For the pentagon, the axes of reflection pass through its vertices and the midpoint of the opposite side.

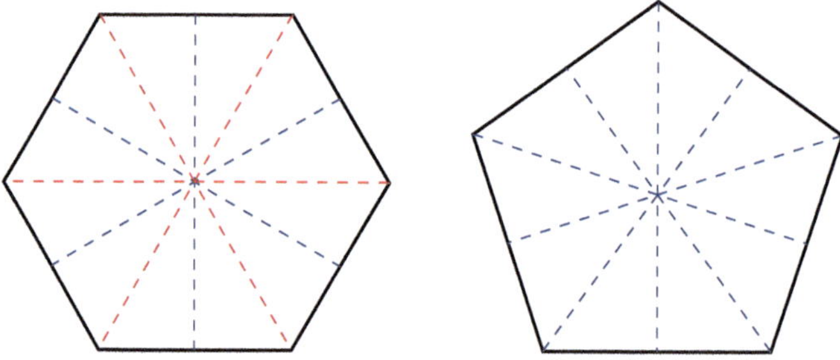

The symmetry group of a regular n-sided polygon is known as the dihedral group $\mathbb{D}_{2n}$. This group consists of n rotations (including the identity rotation) and n reflections, making a total of $2n$ elements.

For the hexagon, the dihedral group $\mathbb{D}_6$ includes 12 elements: 6 rotations and 6 reflections. The rotations are: $\mathbf{D}_{0°}$, $\mathbf{D}_{60°}$, $\mathbf{D}_{120°}$, $\mathbf{D}_{180°}$, $\mathbf{D}_{240°}$, and $\mathbf{D}_{300°}$. The reflections are: $\mathbf{O}_{0°}$, the reflection through the $\vec{x}$-axis, $\mathbf{O}_{60°}$, the reflection across the line through the origin making a counterclockwise angle of $30°$ with the positive $\vec{x}$-axis, $\mathbf{O}_{120°}$, the reflection across the line through the origin making a counterclockwise angle of $60°$ with the positive $\vec{x}$-axis, $\mathbf{O}_{180°}$, $\mathbf{O}_{240°}$, and $\mathbf{O}_{300°}$.

For the pentagon, the dihedral group $\mathbb{D}_{10}$ includes 10 elements: 5 rotations and 5 reflections. The rotations are: $0°, 72°, 144°, 216°$, and $288°$. The reflections originate at each vertex and connect to the midpoint of the opposite side.

Exercise 10.1 Give the formulas for the 10 isometries in $\mathbb{D}_{10}$

Exercise 10.2 How many direct isometries and how many opposite isometries are included in the dihedral group for a regular polygon with 50 sides? with 25 sides?

Exercise 10.3 How does the shape of regular polygons change as the number of sides increases? What shape do they approach? Describe the symmetry group of the this limiting shape.

Since the identity transformation $\mathbf{I} = \mathbf{D}_{0^\circ}$ leaves every every point fixed, every figure in the plane has a symmetry group. If the only isometry in that group is I, the figure is said to be *asymmetric*. Groups, in general, form a very important collection of geometric objects. A group consists of a collection of mathematical objects along with a binary operation that combines two elements in the group to produce another element in the group; for symmetry groups the binary operation is the *composition of isometries*. The basic structure of a group is given by a set of axioms defining how the operation behaves.

The simplest example is the group of integers $\mathbb{Z}$, $\{\ 0, \pm 1, \pm 2, \cdots\ \}$, with addition, $+$, as the binary operation. While we will not delve into the formal axioms here, we use this model to highlight the properties of groups relevant to symmetry groups. In comparing $\mathbb{Z}$ to our symmetry group $\mathbb{D}_{12}$, we first observe that each group has an *identity element*. In $\mathbb{Z}$, the identity element is 0, and its defining property is that $0 + n = n + 0 = n$ for any integer n. Similarly, in our symmetry group, $\mathbf{D}_{0^\circ}$ is the identity element, with the property:

$$\mathbf{D}_{0^\circ} \circ \mathbf{D}_{\theta} = \mathbf{D}_{\theta} \quad \text{and} \quad \mathbf{D}_{\theta} \circ \mathbf{D}_{0^\circ} = \mathbf{D}_{\theta} \quad \text{for all } \theta,$$

as well as

$$\mathbf{D}_{0^\circ} \circ \mathbf{O}_{\theta} = \mathbf{O}_{\theta} \quad \text{and} \quad \mathbf{O}_{\theta} \circ \mathbf{D}_{0^\circ} = \mathbf{O}_{\theta} \quad \text{for all } \theta.$$

The second important property of a group is the existence of inverses. In $\mathbb{Z}$, the inverse of n is $-n$, since $n + (-n) = 0$. In our symmetry group, the inverse of $\mathbf{D}_{\theta}$ is $\mathbf{D}_{360^\circ - \theta}$, as shown by:

$$\mathbf{D}_{360^\circ - \theta} \circ \mathbf{D}_{\theta} = \mathbf{D}_{0^\circ}.$$

Moreover, each opposite isometry is its own inverse:

$$\mathbf{O}_{\theta} \circ \mathbf{O}_{\theta} = \mathbf{D}_{0^\circ}.$$

Exercise 10.4 Since $\mathbb{D}_6$ is a group, the composition of any two isometries in the group is another isometry in the group. To verify, complete filling in this table:

$\circ$	$D_{60°}$	$D_{120°}$	$D_{180°}$	$D_{240°}$	$D_{300°}$	$O_{0°}$	$O_{60°}$	$O_{120°}$	$O_{180°}$	$O_{240°}$	$O_{300°}$
$D_{60°}$	$D_{120°}$										
$D_{120°}$	$D_{180°}$										
$D_{180°}$	$D_{240°}$										
$D_{240°}$	$D_{300°}$										
$D_{300°}$	$D_{0°}$										
$O_{0°}$	$O_{60°}$										
$O_{60°}$	$O_{120°}$										
$O_{120°}$	$O_{180°}$										
$O_{180°}$	$O_{240°}$										
$O_{240°}$	$O_{300°}$										
$O_{300°}$	$O_{80°}$										

Exercise 10.5 Consider the following geometric figures and identify the symmetry group for each. Don't forget the identity will belong to each symmetry group. For each figure, compare the number of direct and opposite isometries.

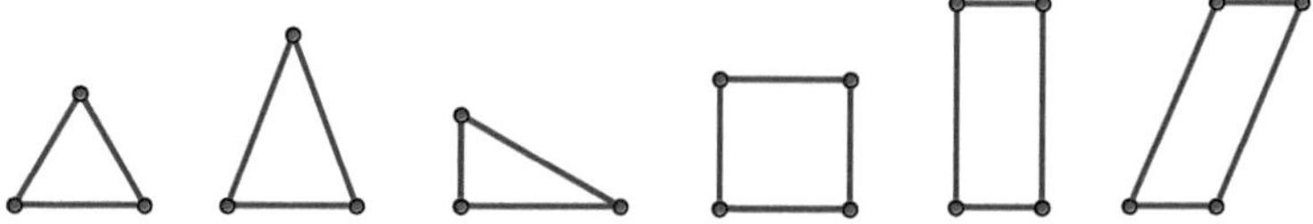

Exercise 10.6 Now consider the symmetry group for each capital letter of the alphabet:

$$A\ B\ C\ D\ E\ F\ G\ H\ I\ J\ K\ L\ M\ N\ O\ P\ Q\ R\ S\ T\ U\ V\ W\ X\ Y\ Z$$

Don't forget the identity will belong to each symmetry group. For each letter, compare the number of direct and opposite isometries.

Considering the geometric figures and letters from the past exercises, you should have observed that these symmetry group either contained no opposite isometries or have an equal number of direct and opposite isometries; remember that the identity isometry, **I**, is a direct isometry.

This is a standard property of planar symmetry groups, which we state with only a sketch of a proof. Suppose there is an opposite isometry **O** and direct isometries $\mathbf{D}_0 = \mathbf{I}, \mathbf{D}_1, \ldots, \mathbf{D}_n$. Then the compositions $\mathbf{O} \circ \mathbf{D}_i$ are all opposite isometries, and they are all distinct. To see this, assume $\mathbf{O} \circ \mathbf{D}_i = \mathbf{O} \circ \mathbf{D}_j$ for $i \neq j$. Composing both sides on the left with **O**, we obtain:

$$\mathbf{O} \circ \mathbf{O} \circ \mathbf{D}_i = \mathbf{O} \circ \mathbf{O} \circ \mathbf{D}_j,$$

and since $\mathbf{O} \circ O = I$, it follows that $\mathbf{D}_i = \mathbf{D}_j$, contradicting $i \neq j$.

Could we have missed an opposite isometry? Let $\mathbf{P}$ be any opposite isometry. Then consider $\mathbf{O} \circ P$. This must be a direct isometry, so $\mathbf{O} \circ \mathbf{P} = \mathbf{D}_i$ for some i. Composing both sides of this equality with $\mathbf{O}$ yields:

$$\mathbf{P} = \mathbf{O} \circ \mathbf{O} \circ \mathbf{P} = \mathbf{O} \circ \mathbf{D}_i.$$

Thus, every opposite isometry can be expressed as $\mathbf{O} \circ \mathbf{D}_i$, confirming that we have not missed any opposite isometries.

In Exercise 10.3, you discovered the symmetry group of the circle. In Exercise 10.7, you will extend that investigation. Before doing so, let us explore the symmetry group of a line ℓ. Let $\mathbb{S}_\ell$ denote the set of isometries of the plane that fix ℓ. Recall that for any $L \in \mathbb{S}_\ell$, the restriction of L to ℓ is an isometry of ℓ.

To simplify our notation, let ℓ be the x-axis. From Part 1, we know that the isometries of ℓ are either translations or reflections over a fixed point. Let us examine each of these cases and determine which isometries of the plane agree with these one-dimensional isometries.

First, consider a translation by the vector $\vec{w} = \lambda \vec{x}$ on the x-axis. Then the isometries $L = \mathbf{D}_{0°} + \vec{w}$ and $K = \mathbf{O}_{0°} + \vec{w}$ both fix the x-axis and translate it by $\vec{w}$.

Now, consider the isometry of ℓ that reflects the line over the point $\vec{c}$. In this case, the isometries $H = \mathbf{D}_{180°} + 2\vec{c}$ and $J = \mathbf{O}_{180°} + 2\vec{c}$ fix the x-axis and correspond to this one-dimensional isometry of the line.

Here, we again observe a one-to-one correspondence between the direct and opposite isometries of the symmetry group of the line.

Exercise 10.7 Compute the isometries in the symmetry group of the unit circle centered at the origin.

10.2 Frieze Groups

One-dimensional repeating patterns, such as decorative borders in art and architecture, are described mathematically by frieze groups.

Specifically, a frieze pattern is an infinite pattern that has translational symmetry; that is, the pattern can be shifted to the right or left by integer multiples of a fixed distance and remain unchanged. In addition to translation, a pattern may exhibit other symmetries. The set of all symmetries of a frieze pattern forms a frieze group.

We conclude this book with a final set of exercises leading to a research project on the frieze groups.

Exercise 10.8 A frieze pattern may have additional symmetries beyond translation. Below are two strips. For each strip, describe all additional symmetries that you can find.

Strip 1

$$\underline{\overline{\Sigma Z \Sigma Z \Sigma Z \Sigma Z \Sigma Z \Sigma Z \Sigma Z \Sigma}}$$

Strip 2

$$\underline{\overline{\mathsf{X\,X\,X\,X\,X\,X\,X\,X\,X\,X\,X}}}$$

Exercise 10.9 Make a complete list of the simplest possible frieze symmetries. For example, an infinite set of equally spaced vertical lines of reflection. Since the composition of two adjacent vertical reflections is the translation, it is automatically included. (The answers are included below.)

Exercise 10.10 For each of these basic isometries construct a strip that includes none of the others, except for translation

a. Translations only:

b. Parallel Reflection and Translations only:

c. Perpendicular Reflections and Translations only:

d. Half-turns and Translations only:

e. Primitive Glide Reflections and Translations only:

Exercise 10.11 How do these symmetries interact? For each case below, construct a frieze pattern with the listed symmetries. What other symmetries are you forced to include?

a. Parallel Reflection and Half-turns:

b. Primitive Glide Reflection and Half-turns:

Project 10.2 As a research project, determine all possible symmetry groups of frieze patterns and to construct a representative example of each.

Matrices

A

A.1   Basis Definitions

Matrices are one of the most important tools of linear algebra that we need to describe 2D linear functions. In this appendix, we develop this tool. By an $n \times m$ *matrix* we mean an $n \times m$ array of real numbers: n rows of length m (called *row vectors*) or m columns of height n (called *column vectors*). Rows are indexed top to bottom from 1 to n and columns from left to right from 1 to m. We denote matrices by bold capital letters: an $n \times m$ matrix is often written $\mathbf{M}_{(n,m)}$, or simply $\mathbf{M}$ when the dimensions are understood. The entry in the ith row and j^{th} column of $\mathbf{M}$ is $\mathbf{M}_{ij}$.

Matrix Operations We have three basic algebraic operations on matrices:

(1) **Scalar Multiplication.** Given an $n \times m$ matrix $\mathbf{M}$ and a scalar $s \in \mathbb{R}$, the matrix $s\mathbf{M}$ is defined by

$$(s\mathbf{M})_{ij} = s\,\mathbf{M}_{ij}, \qquad 1 \le i \le n,\ 1 \le j \le m.$$

(2) **Matrix Addition.** If $\mathbf{M}$ and $\mathbf{N}$ are both $n \times m$ matrices, then their sum $\mathbf{M} + \mathbf{N}$ is defined entrywise by

$$(\mathbf{M} + \mathbf{N})_{ij} = \mathbf{M}_{ij} + \mathbf{N}_{ij}.$$

(3) **Matrix Multiplication.** If $\mathbf{M}$ is $n \times k$ and $\mathbf{N}$ is $k \times m$, then the product $\mathbf{MN}$ is the $n \times m$ matrix defined by

© The Author(s), under exclusive license to Springer Nature Switzerland AG 2026
J. J. Edmond, J. E. Graver, *The Linear Function and Euclidean Geometry*, Synthesis
Lectures on Mathematics & Statistics, https://doi.org/10.1007/978-3-032-22712-6

$$(\mathbf{MN})_{ij} = \sum_{h=1}^{k} \mathbf{M}_{ih}\, \mathbf{N}_{hj}, \qquad 1 \le i \le n,\ 1 \le j \le m.$$

Special Matrices

(1) The *identity matrix* $\mathbf{I}_n$ is the $n \times n$ matrix

$$(\mathbf{I}_n)_{ij} = \begin{cases} 1, & i = j, \\ 0, & i \ne j. \end{cases}$$

(2) The *transpose* of an $n \times m$ matrix $\mathbf{M}$ is the $m \times n$ matrix $\mathbf{M}^T$ defined by

$$(\mathbf{M}^T)_{ij} = \mathbf{M}_{ji}.$$

(3) An $n \times n$ matrix $\mathbf{M}$ is *invertible* if there exists an $n \times n$ matrix $\mathbf{M}^{-1}$ such that

$$\mathbf{M}^{-1}\mathbf{M} = \mathbf{I}_n = \mathbf{M}\mathbf{M}^{-1}.$$

Exercise A.1 Compute $\mathbf{MM}^T$, $\mathbf{NM}$, and $\mathbf{N}^T\mathbf{N}$ where

$$\mathbf{M} = \begin{bmatrix} 1 & 1 & 1 \\ 1 & 1 & 0 \\ 0 & 1 & 1 \end{bmatrix}, \qquad \mathbf{N} = \begin{bmatrix} 1 & 1 & 1 \\ 1 & 0 & 1 \\ 1 & 1 & 0 \\ 0 & 1 & 1 \end{bmatrix}.$$

Proposition A.1 *Let $\mathbf{L}, \mathbf{M}, \mathbf{N}$ be $m \times n$ matrices and let r, s be scalars. Then:*

(1) $\mathbf{M} + \mathbf{N} = \mathbf{N} + \mathbf{M}$ *(commutativity)*
(2) $(\mathbf{L} + \mathbf{M}) + \mathbf{N} = \mathbf{L} + (\mathbf{M} + \mathbf{N})$ *(associativity)*
(3) The zero matrix $\mathbf{0}$ is the additive identity: $\mathbf{M} + \mathbf{0} = \mathbf{M}$
(4) Every matrix $\mathbf{M}$ has an additive inverse $-\mathbf{M}$: $\mathbf{M} + (-\mathbf{M}) = \mathbf{0}$
(5) $s(\mathbf{M} + \mathbf{N}) = s\mathbf{M} + s\mathbf{N}$
(6) $(r + s)\mathbf{M} = r\mathbf{M} + s\mathbf{M}$
(7) $(rs)\mathbf{M} = r(s\mathbf{M})$
(8) $1 \cdot \mathbf{M} = \mathbf{M}$

Proof Each identity follows from the corresponding identity for real numbers applied entrywise. For example,

$$((\mathbf{L} + \mathbf{M}) + \mathbf{N})_{ij} = \mathbf{L}_{ij} + \mathbf{M}_{ij} + \mathbf{N}_{ij} = (\mathbf{L} + (\mathbf{M} + \mathbf{N}))_{ij}.$$

The other statements follow similarly. $\square$

Proposition A.2 *Let $s, t \in \mathbb{R}$ and let $\mathbf{L}$ be an $h \times k$ matrix, $\mathbf{M}$ and $\mathbf{M}'$ be $k \times n$ matrices, and $\mathbf{N}$ be an $n \times m$ matrix. Then:*

(1) $(\mathbf{LM})\mathbf{N} = \mathbf{L}(\mathbf{MN})$ *(associativity)*
(2) $(st)\mathbf{M} = s(t\mathbf{M})$ *and* $(s\mathbf{M})\mathbf{N} = s(\mathbf{MN})$
(3) $\mathbf{I}_k\mathbf{M} = \mathbf{M} = \mathbf{MI}_n$
(4) $\mathbf{L}(\mathbf{M} + \mathbf{M}') = \mathbf{LM} + \mathbf{LM}'$ *and* $(\mathbf{M} + \mathbf{M}')\mathbf{N} = \mathbf{MN} + \mathbf{M}'\mathbf{N}$

Exercise A.2 Let $\mathbf{M}$ be $k \times n$ and $\mathbf{N}$ be $n \times m$. Compute $(\mathbf{MN})_{ij}^{T}$ and $(\mathbf{N}^T\mathbf{M}^T)_{ij}$ to show $(\mathbf{MN})^{T} = \mathbf{N}^T\mathbf{M}^T$.

A.2   Matrices as Functions and Determinants

We may identify $\mathbb{R}^n$ with the space of $n \times 1$ matrices, or equivalently, with the space of n-dimensional column vectors. Let $\mathbf{M}$ be an $n \times m$ matrix and let $\vec{v}$ be an $m \times 1$ column vector. Then the matrix product yields

$$\mathbf{M}\vec{v} = \vec{u},$$

where $\vec{u}$ is an $n \times 1$ column vector. Thus, an $n \times m$ matrix $\mathbf{M}$ may be viewed as a function from $\mathbb{R}^m$ to $\mathbb{R}^n$.

Our primary interest will be in 2×2 matrices that map $\mathbb{R}^2$ onto $\mathbb{R}^2$, that is, linear transformations of the plane.

Consider the 2×2 matrix

$$\mathbf{M} = \begin{bmatrix} a & c \\ b & d \end{bmatrix}$$

and the mapping $L(\vec{v}) = \mathbf{M}\vec{v}$. We define the **determinant** of $\mathbf{M}$ to be $ad - bc$, which we denote by $\Delta(\mathbf{M})$, or simply Δ. We say that $\mathbf{M}$ is *nonsingular* if $\Delta(\mathbf{M}) \neq 0$. All of the matrices considered in this text will be nonsingular.

Let $\mathbf{M}$ be nonsingular. Then:

(1) The transformation L is one-to-one and onto.
(2) The transformation L has an inverse.
(3) The matrix of L^{-1} is $\mathbf{M}^{-1}$ (the inverse of $\mathbf{M}$).

Next we would like to understand how the matrix $\mathbf{M}$ changes distances and areas. Consider the matrix $\mathbf{M} = \begin{bmatrix} 2 & 1 \\ -\frac{1}{2} & \frac{3}{2} \end{bmatrix}$. We see that $\mathbf{M}$ maps the x–unit vector (the first standard basis vector) to $\begin{bmatrix} 2 \\ -\frac{1}{2} \end{bmatrix}$ and the y–unit vector (the second basis vector) to $\begin{bmatrix} 1 \\ \frac{3}{2} \end{bmatrix}$. Then, the unit square with vertices $\begin{bmatrix} 0 \\ 0 \end{bmatrix}, \begin{bmatrix} 0 \\ 1 \end{bmatrix}, \begin{bmatrix} 1 \\ 1 \end{bmatrix}, \begin{bmatrix} 1 \\ 0 \end{bmatrix}$ is mapped onto the parallelogram with vertices $\begin{bmatrix} 0 \\ 0 \end{bmatrix}, \begin{bmatrix} 1 \\ \frac{3}{2} \end{bmatrix}, \begin{bmatrix} 3 \\ 1 \end{bmatrix}, \begin{bmatrix} 2 \\ -\frac{1}{2} \end{bmatrix}$. We also note that the matrix $\mathbf{N} = \begin{bmatrix} 1 & 2 \\ \frac{3}{2} & -\frac{1}{2} \end{bmatrix}$ maps the x–unit vector onto $\vec{u}$ and the y–unit vector onto $\vec{v}$. Again, the unit square is mapped onto the same parallelogram; however, the clockwise orientation around the unit square is mapped onto a counterclockwise orientation of the parallelogram.

Proposition A.3 *Let $\mathbf{M}$ be a real nonsingular 2×2 matrix. Then:*

(1) For any region R, $\text{area}(\mathbf{M}(R)) = \det(\mathbf{M}) \cdot \text{area}(R)$.
(2) If $\det(\mathbf{M}) > 0$, all orientations are preserved; if $\det(\mathbf{M}) < 0$, all orientations are reversed.

We will not give a formal proof of this proposition. Instead, we explain the underlying geometric idea. Consider the matrix transformation $L(\vec{v}) = \mathbf{M}\vec{v}$, where $\mathbf{M} = \begin{bmatrix} a & c \\ b & d \end{bmatrix}$. Below we visualize this transformation.

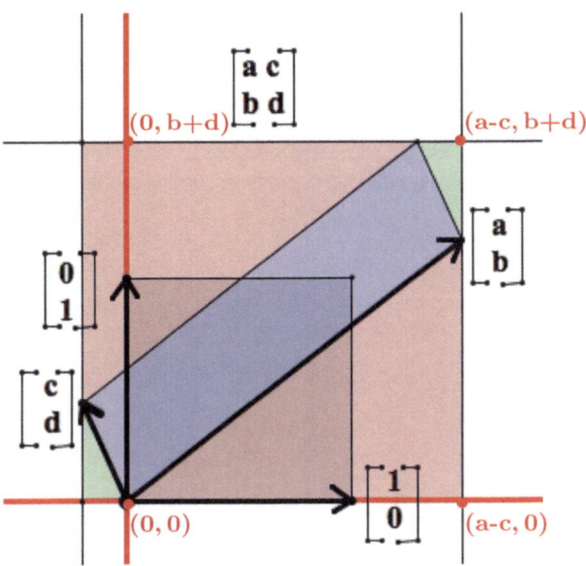

The image of the x unit vector, $\begin{bmatrix} 1 \\ 0 \end{bmatrix}$, is mapped onto $\begin{bmatrix} a \\ b \end{bmatrix}$ and the y unit vector, $\begin{bmatrix} 0 \\ 1 \end{bmatrix}$, is mapped onto $\begin{bmatrix} c \\ d \end{bmatrix}$ Thus T maps the unit square (shown in gray) onto the parallelogram spanned by the vectors $\begin{bmatrix} a \\ b \end{bmatrix}$ and $\begin{bmatrix} c \\ d \end{bmatrix}$ (shown in blue). We assert that the area of this parallelogram is $ad - bc$, the determinant of $\mathbf{M}$.

To compute the area, we enclose the parallelogram in an axis–aligned rectangle (shown in red) and subtract the areas of the triangular regions outside the parallelogram. Because c is negative in this picture, the width of the rectangle is $a - c$ and the height is $b + d$. Thus its area of the rectangle is $(a - c)(b + d)$. Each of the two pink triangles has area $\frac{1}{2}ab$, so their total area is ab. Similarly, the two green triangles contribute $(-c)d$. Thus the area of the parallelogram is
$$(a - c)(b + d) - ab - (-c)d - (ab + ad - cb - cd) - ab + cd - ad - bc$$
It follows that when the determinant is 0, this parallelogram has area 0 and collapses to a line segment: both vectors $\begin{bmatrix} a \\ b \end{bmatrix}$ and $\begin{bmatrix} c \\ d \end{bmatrix}$ lie on the same line; explaining why we restrict out attention to non-singular matrix transformations.

Index

J. J. Edmond, J. E. Graver, *The Linear Function and Euclidean Geometry*, Synthesis
Lectures on Mathematics & Statistics, https://doi.org/10.1007/978-3-032-22712-6